AF588224

Synthesis Lectures on Emerging Engineering Technologies

This series publishes short books on current engineering technologies that are gaining prominence, as well as promising technologies that are being developed, for an audience of researchers, advanced students, engineers and other professionals, and entrepreneurs.

Ibrahim Yousef • Sirish L. Shah •
R. Bhushan Gopaluni

Visual Analytics for Process Monitoring

From Time Series to Interpretable Visual Representations

Springer

Ibrahim Yousef
Chemical and Biological Engineering
University of British Columbia
Vancouver, BC, Canada

Sirish L. Shah
Chemical and Materials Engineering
University of Alberta
Edmonton, AB, Canada

R. Bhushan Gopaluni
Chemical and Biological Engineering
University of British Columbia
Vancouver, BC, Canada

ISSN 2381-1412 ISSN 2381-1439 (electronic)
Synthesis Lectures on Emerging Engineering Technologies
ISBN 978-3-032-22125-4 ISBN 978-3-032-22126-1 (eBook)
https://doi.org/10.1007/978-3-032-22126-1

This Springer imprint is published by the registered company Springer Nature Switzerland AG
The registered company address is: Gewerbestrasse 11, 6330 Cham, Switzerland

Preface

This monograph presents a revised and compact version of my doctoral dissertation completed at the University of British Columbia. The content reflects the culmination of my PhD research conducted at the Data Analytics and Intelligent Systems (DAIS) lab.

During my PhD, I was fortunate to collaborate with both industrial partners and academic peers who provided valuable feedback and data access that helped shape this research. The monograph consolidates the key ideas and findings of my research into a compact and coherent narrative.

All research presented here was carried out under my primary authorship, with responsibilities including data analysis, methodology development, and manuscript writing. Dr. R. Bhushan Gopaluni and Dr. Sirish L. Shah were the supervisory authors, involved in conceptual development and manuscript editing. Other collaborators contributed to concept formation, data acquisition, and/or manuscript editing.

This work draws upon the following published research outputs, which collectively form the foundation of this monograph:

- Lee D. Rippon, **Ibrahim Yousef**, Behrooz Hosseini, Arbi Bouchoucha, Jean-Francois Beaulieu, Carole Prévost, Michel Ruel, Sirish L. Shah, and R. Bhushan Gopaluni. Representation learning and predictive classification: Application with an electric arc furnace. *Computers & Chemical Engineering*, 150:107304, 2021.
- **Ibrahim Yousef**, Sirish L. Shah, and R. Bhushan Gopaluni. Visual Analytics: A New Paradigm for Process Monitoring. *IFAC-PapersOnLine*, 55(7), 2022.
- **Ibrahim Yousef**, Aditya Tulsyan, Sirish L. Shah, and R. Bhushan Gopaluni. Visual Analytics for Process Monitoring: Leveraging Time-Series Imaging for Enhanced Interpretability. *Journal of Process Control*, 132:103127, 2023.
- **Ibrahim Yousef**, Lee D. Rippon, Carole Prévost, Sirish L. Shah, and R. Bhushan Gopaluni. The arc loss challenge: A novel industrial benchmark for process analytics and machine learning. *Journal of Process Control*, 128:103023, 2023.
- **Ibrahim Yousef**, Sirish L. Shah, and R. Bhushan Gopaluni. Time Series Representation Learning via Cross-Domain Predictive and Contextual Contrasting: Application to Fault Detection. *Engineering Applications of Artificial Intelligence*, 0952-1976, 2025.

Permission to reproduce figures and materials from these publications has been obtained from the respective publishers.

I hope this monograph serves as a useful reference for researchers and practitioners in the fields of process systems engineering, data analytics, and machine learning, particularly those interested in interpretable approaches to industrial process monitoring.

Acknowledgments Ibrahim Yousef would like to acknowledge the collaborators at BBA Engineering Consultants for their valuable industrial insights, and his research collaborators, especially Dr. Lee D. Rippon and the members of the DAIS Lab, for their support and engaging discussions. Finally, he is deeply grateful to his family and friends for their constant encouragement and unconditional support, without which this work would not have been possible.

Sirish L. Shah would like to acknowledge funding support from the Natural Sciences and Engineering Research Council of Canada (NSERC) for this work. The drive to explore, discover and gain insight is the art of creation. He is thankful to his graduate students, research fellows, colleagues, and visiting fellows who were all drawn to this art of research and have been his gifted teachers. He is deeply grateful to his wife, children and grandchildren for their unconditional support.

Bhushan Gopaluni would like to acknowledge the generous funding from the NSERC and Mitacs, without which much of the research underlying this book would not have been possible. He would also like to thank the many graduate students he has had the privilege of supervising over the years. Their curiosity, dedication, and insight have all contributed to the ideas in this book, either directly or indirectly. Finally, he thanks his family—Vidya, Syon, and Myra—for their unwavering support and patience.

Vancouver, BC, Canada
December 2025

Ibrahim Yousef

Declarations

Competing Interests The authors have no competing interests to declare that are relevant to the content of this manuscript.

Ethics Approval No ethics approval was required.

Contents

Acronyms

ACC	Accuracy
ANN	Artificial Neural Network
BOSS	Bag of Symbolic-Fourier-Approximation Symbols
CDPCC	Cross-Domain Predictive and Contextual Contrasting
CNN	Convolutional Neural Network
CPC	Contrastive Predictive Coding
CSTH	Continuous Stirred Tank Heater
CW	Cold Water
DC EAF	Direct Current Electric Arc Furnace
DTW	Dynamic Time Warping
FCN	Fully Connected Network
FDD	Fault Detection and Diagnosis
FF	Furnace Feed
FFT	Fast Fourier Transform
FOGET	Furnace Off-Gas Temperature
FPR	False Positive Rate
GAF	Gramian Angular Field
GRU	Gated Recurrent Unit
HW	Hot Water
LLM	Large Language Models
LSTM	Long Short-Term Memory
ML	Machine Learning
MLP	Multi-Layer Perceptron
MTS	Multivariate Time Series
NLP	Natural Language Processing
PAA	Piecewise Aggregate Approximation
PCA	Principal Component Analysis
PLS	Partial Least Squares
PPV	Positive Predictive Value (= Precision)
PSE	Process Systems Engineering
RP	Recurrence Plot

RNN	Recurrent Neural Network
SFA	Symbolic Fourier Approximation
SimCLR	Simple Framework for Contrastive Learning of Representations
SPE	Squared Prediction Error
SSL	Self-Supervised Learning
TCC	Temporal Contextual Contrasting
TEP	Tennessee Eastman Process
T-FC	Time-Frequency Consistency
TP	Total Power
TPR	True Positive Rate (= Recall)
TSC	Time Series Classification
TSF	Time Series Forest
TT	Training Time

Introduction to Visual Analytics

1

Contents

1.1 Evolution of Process Monitoring

Compared to previous decades, modern industrial systems generate and collect unprecedented volumes of data. This is driven by the widespread deployment of high-speed sensors, integrated databases, and advanced computing infrastructure (e.g., cloud and high-performance computing) [1]. Industrial data volumes are now doubling approximately every two years [2]. In 2021, industries worldwide had collected and consumed an estimated 74 zettabytes (74×10^{21} bytes) of digital data [3]. In contrast, a few decades ago, data collection was limited to periodic manual readings or low-frequency samples stored locally. Today, virtually every process parameter can be measured in real time, producing rich data streams at high sampling rates. This rapid growth of data is a defining feature of the Industry 4.0 era, which emphasizes digital connectivity and smart automation across manufacturing and process industries [4].

The availability of high-frequency, high-dimensional data has opened opportunities across a wide range of industries, including the chemical process industry. Among various industrial applications, process monitoring has particularly benefited from this abundance of data. Process monitoring involves the continuous observation of process variables to detect, diagnose, and respond to abnormal conditions, commonly referred to as faults [5]. A fault is typically defined as an unpermitted deviation of at least one process parameter from its normal operating condition, potentially leading to degraded performance, safety

I. Yousef et al., *Visual Analytics for Process Monitoring*, Synthesis Lectures on Emerging Engineering Technologies, https://doi.org/10.1007/978-3-032-22126-1_1

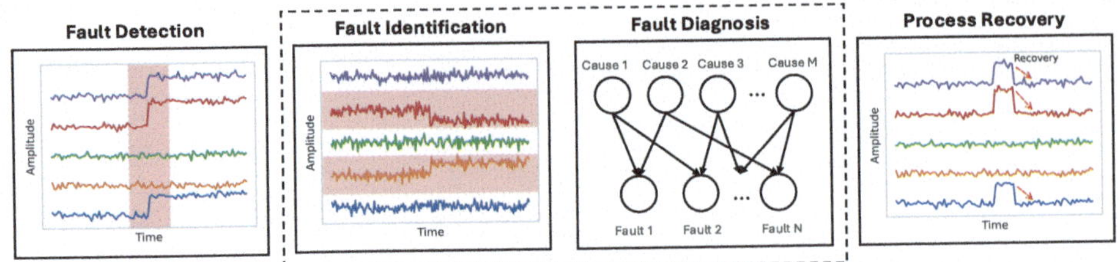

Fig. 1.1 The four main components of process monitoring: (1) fault detection: detecting abnormal behavior, (2) fault identification: identifying affected variables, (3) fault diagnosis: inferring the root cause, and (4) process recovery: returning the system to normal operation

risks, or equipment damage [6, 7]. The process monitoring framework generally includes four key components: fault detection, which determines whether a fault has occurred; fault identification, which identifies the variables that are most relevant to the fault; fault diagnosis, which identifies the root cause of the fault; and process recovery, which focuses on eliminating the effect of the fault on the process [5, 8]. Figure 1.1 provides a visual summary of the four components of process monitoring. In practice, fault identification and fault diagnosis are often addressed jointly, as identifying the affected variables typically supports the inference of the underlying cause. In this monograph, we use the term fault diagnosis to refer to both steps.The scope of this monograph is limited to fault detection and diagnosis (FDD).

Historically, process monitoring relied heavily on first-principles models, rule-based techniques, and expert knowledge [9]. For instance, control charts and simple threshold-based alarms (i.e., fault trees) were widely used to detect faults in individual process variables. While interpretable and easy to implement, such methods are often designed for single-variable monitoring [10]. The next category of methods introduced multivariate statistical techniques such as principal component analysis (PCA) and partial least squares (PLS) [11]. This category focuses on the simultaneous monitoring of multiple correlated variables by projecting high-dimensional process data into a lower-dimensional latent space that captures the dominant variance in the original space [12]. The continued expansion of sensor networks and computing power paved the way for a third category of learning-based methods, including machine learning and deep learning models [13]. The promise of these methods lies in their ability to learn complex patterns directly from high-dimensional and time-varying processes [14]. As a result, they can capture subtle and nonlinear faults with minimal reliance on prior process knowledge. Consequently, this category has attracted growing interest in both academic research and industrial applications in recent years. Figure 1.2 illustrates the three main categories of process monitoring approaches.

Despite the great promise shown by modern data-driven approaches in the literature [15, 16], several challenges still limit their widespread deployment for FDD in industrial settings. One key limitation is the lack of interpretability associated with complex models [17–19]. In high-risk industrial settings, domain experts require clear reasoning behind

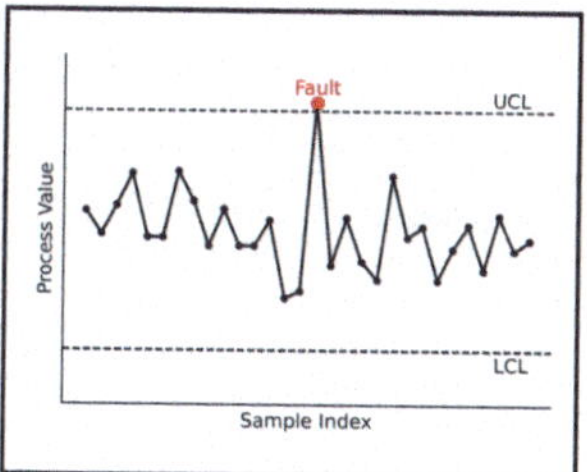

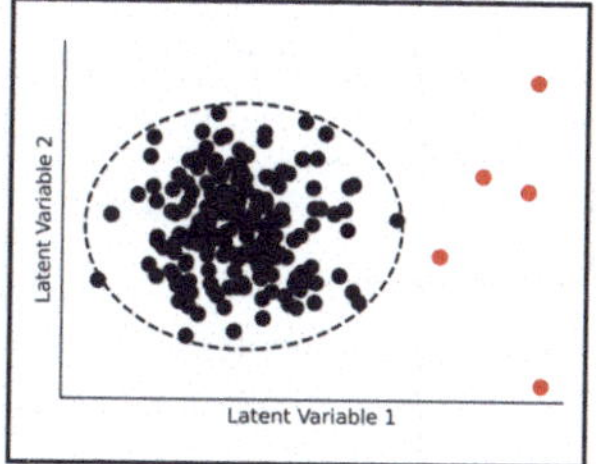

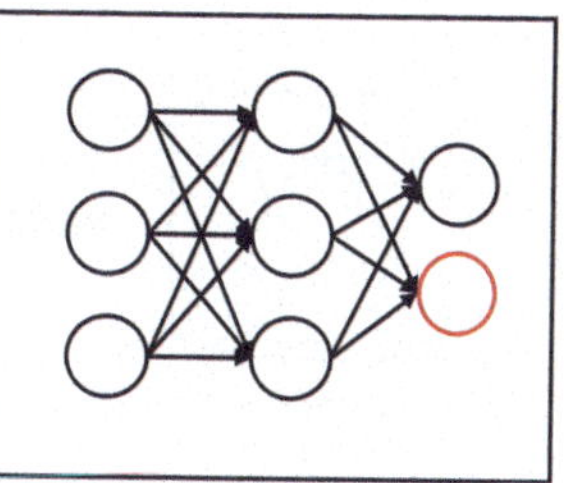

Time

Fig. 1.2 An illustration of the evolution of process monitoring techniques. The left panel represents early model-based monitoring approaches using univariate control charts for detecting faults in individual variables. The middle panel illustrates multivariate statistical process monitoring, where methods like PCA identify faults by projecting high-dimensional data into a latent space. The right panel shows modern data-driven approaches, where deep neural networks learn complex patterns from process data for FDD

model predictions to trust and act upon model-generated alarms. Next, supervised learning dominates modern FDD research, which requires all process data for each time interval to be labelled with the corresponding process state. This constraint poses a major bottleneck, as manual labelling is time-consuming, error-prone, and labour-intensive. Moreover, acquiring high-quality data that contain recurring and clearly identifiable fault patterns in industrial environments is inherently difficult due to the rarity, variability, and unpredictable nature of real-world faults [20]. Thirdly, most of the academic work in FDD is benchmarked using simulated datasets under idealized conditions, such as the Tennessee Eastman Process (TEP) [21]. Simulated datasets often lack critical challenges such as high dimensionality, class imbalance, and noise, which are inherent in practical industrial scenarios. Therefore, many proposed methods lack robustness and underperform when deployed in practice [22].

1.2 Problem Formulation

Data-driven FDD methods are commonly classified into three main categories: supervised, unsupervised, and semi-supervised, depending on the availability of faulty data in the training set. In the supervised category, the FDD task is formulated as a classification problem, where the model is trained on labelled datasets that include sufficient samples from both normal and faulty classes [23]. The objective is to learn discriminative features associated with each class. In contrast, unsupervised approaches assume that only data from normal operating conditions are available during training [24]. As shown in Fig. 1.3, the model is first trained offline using normal samples to compute a monitoring statistic. Once training is complete, a detection threshold δ is determined based on the distribution of this statistic across the training data. During online monitoring, the trained

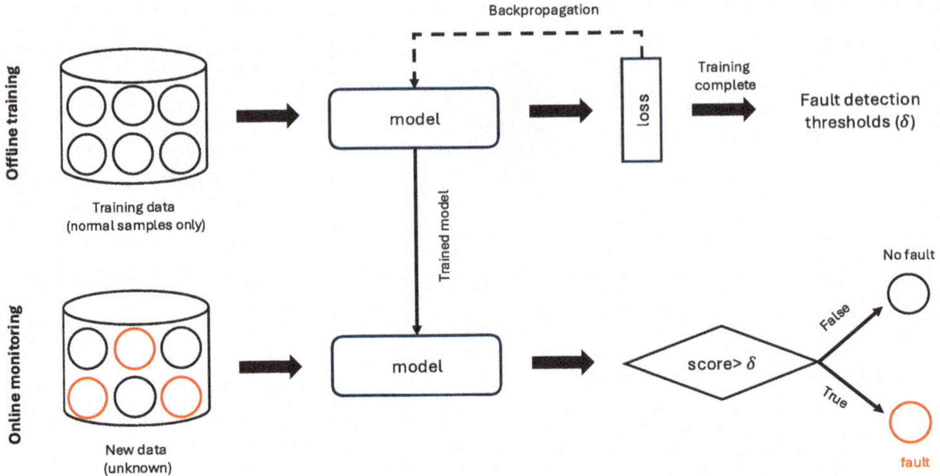

Fig. 1.3 A high-level framework of the unsupervised category for FDD. During the offline training phase, the model is trained using only normal samples to learn baseline (normal) system behaviour. A fault detection threshold δ is then computed based on the distribution of a monitoring statistic. Commonly used statistics include the squared prediction error (SPE) in PCA-based methods [25] and reconstruction error in autoencoder architectures [24]. In the online monitoring phase, incoming samples are evaluated using the trained model. If the computed score exceeds δ, the sample is classified as faulty; otherwise, it is considered normal

model computes the monitoring statistic for each incoming sample and compares it with δ. Samples exceeding δ are classified as faulty, while those below the threshold are considered normal. Lastly, semi-supervised approaches address scenarios where labelled faulty data are limited but large amounts of unlabeled or normal data are available [20]. The focus of this monograph is on the supervised category of FDD approaches, where we assume that sufficient labelled faulty samples are available in the training set.

Industrial systems are instrumented with a network of sensors that continuously record process variables across multiple dimensions and over time. These measurements capture the dynamic behaviour of the system and reflect changes in operating conditions. A univariate time series signal, denoted as $x_i = \{x_{i,1}, x_{i,2}, \ldots, x_{i,L}\}$, represents a sequence of L chronologically ordered measurements recorded over time. Each measurement is associated with a timestamp from the set $T = \{t_1, t_2, .., t_L\}$. When p different time series signals are recorded simultaneously by a set of p sensors, the data are referred to as a multivariate time series (MTS) signal, denoted as $X_i = \{x_1, x_2, \ldots, x_p\}$, where $x_j \in \mathbb{R}^L$. We assume synchronized and fixed sampling across all dimensions and omit explicit time indexing in the MTS notation for simplicity.

In the supervised set-up, we work with labelled MTS datasets, denoted by $\mathcal{D} = \{(X_1, Y_1), (X_2, Y_2), \ldots, (X_N, Y_N)\}$, which contain a collection of N paired samples (X_i, Y_i). Each sample consists of an MTS signal X_i and its corresponding label Y_i. Figure 1.4 presents the high-level representation of a supervised FDD framework. The

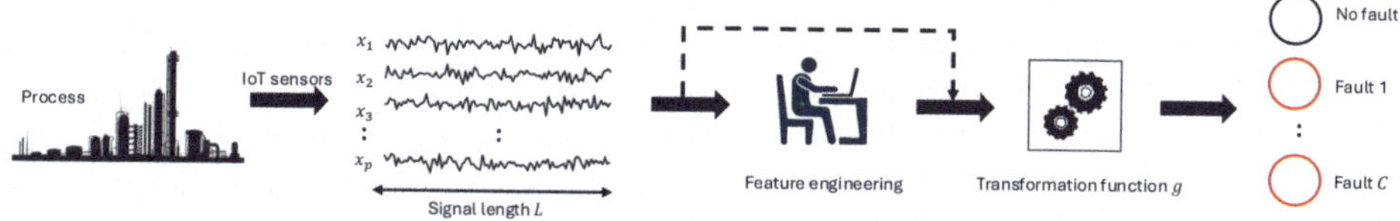

Fig. 1.4 An illustration of the supervised FDD category. MTS data from IoT sensors are either manually transformed via feature engineering or directly fed into a model g. The model g maps inputs, or their feature representations, to class labels representing the underlying operating conditions

focus of this is on training a model g that learns a mapping from the input X_i (or its feature representations) to its actual label Y_i (i.e., $g : X \longrightarrow Y$). Here, the labels represent the operating condition of a system. For one-fault classification problems, $Y_i = 0$ denotes the normal condition, and $Y_i = 1$ denotes the faulty condition. In multi-fault classification problems, each sample is associated with a class label $Y_i \in \{0, 1, \ldots, C\}$ where $C - 1$ represents the total number of distinct fault categories.

1.3 Visual Analytics: Motivation and Definition

This section introduces visual analytics in the context of process monitoring and FDD. Visual analytics maps process data into visual representations, enabling the application of image-based models to analyze complex patterns while allowing process experts to intuitively interact with the data. The following subsections present the motivation for adopting visual analytics, followed by its formal definition and methodological framework.

1.3.1 Motivation

FDD problems, formulated as time series classification (TSC) problems, present unique challenges compared to traditional supervised learning on tabular data. Specifically, process data exhibit strong temporal dependencies (i.e., autocorrelations), where the order of measurement carries valuable information. These dynamics violate the I.I.D. (independent and identically distributed) assumption foundational to many machine learning models [26]. Furthermore, the state-of-the-art time series classifiers have a high computational cost and often scale poorly for longer series and many samples. For example, the computational complexity of HIVE-COTE 2.0 (Hierarchical Vote Collective of Transformation-based Ensembles V2), one of the best-performing time series classifier on benchmark time series archives, evolves approximately according to a power-law relationship with respect to the time series length L, the number of training samples N, and the number of variables p [27]. This complexity limits the deployment of state-of-the-art time series classifiers (most of which are ensemble-based) in industrial settings where fast predictions are necessary.

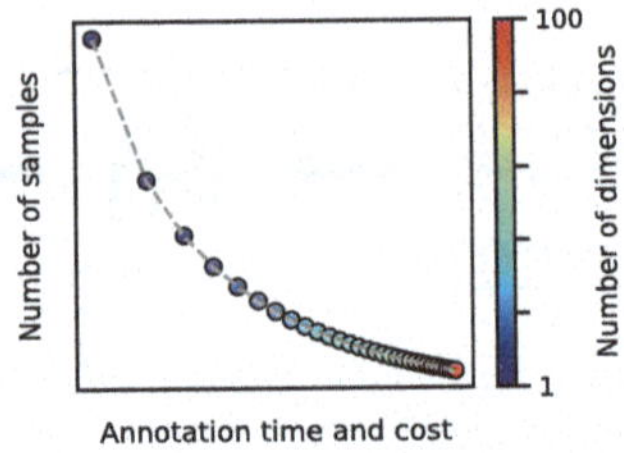

Fig. 1.5 A conceptual illustration of the trade-off between annotation time and cost and the number of labelled samples as system dimensionality increases. In low-dimensional systems, a large number of samples can be annotated with minimal effort. However, as dimensionality increases, the time and expertise required to label the data grow substantially, resulting in fewer annotated samples despite higher annotation effort

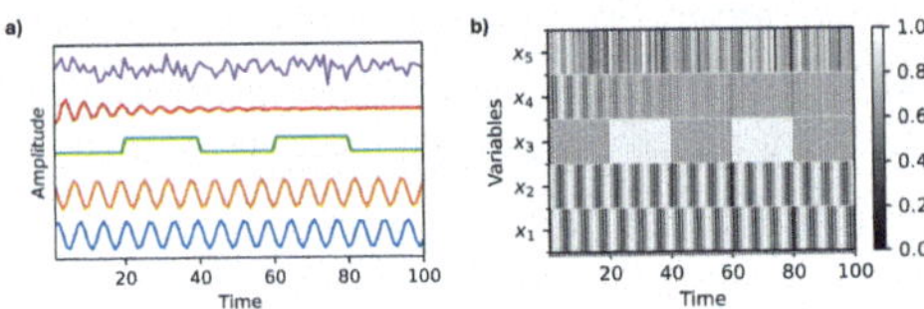

Fig. 1.6 A visual representation of MTS as grayscale images. (**a**) Simulated signals from five process variables in the time domain. (**b**) The same MTS reshaped into a grayscale image, where time defines the width, variables define the height, and pixel intensity represents the signal amplitude

Next, annotating faults in high-dimensional systems, assuming faults are known, is time-consuming and labour-intensive (cf. Fig. 1.5). Faulty events are typically short-lived, subtle, and obscured by noise [28]. This motivates the development of workflows that provide process experts with intuitive tools that facilitate better interaction with the data.

Interestingly, there are striking conceptual similarities between FDD and classical computer vision tasks such as image classification [29]. In the latter, successful algorithms must learn from the spatial information contained in images. Similarly, the core problem in FDD is fundamentally the same, albeit with one less dimension [30]. For example, process data (or MTS data) can be represented as grayscale images, where the number of variables p corresponds to the image height, the number of time steps L forms the width, and the signal amplitude is encoded as a single-channel pixel intensity. Figure 1.6 illustrates this analogy; it shows how process data can be treated as a grayscale image. This analogy is compelling because it highlights the potential to transfer methods from computer vision to the FDD domain. For instance, convolutional neural networks (CNN), commonly used in computer vision, have been successfully adapted to capture temporal dependencies and extract patterns from process data [31–33].

Despite the promise of this cross-domain transfer, the differences between image and process data present unique challenges [34]. In image data, each pixel is typically represented by a three-dimensional vector corresponding to the red, green, and blue

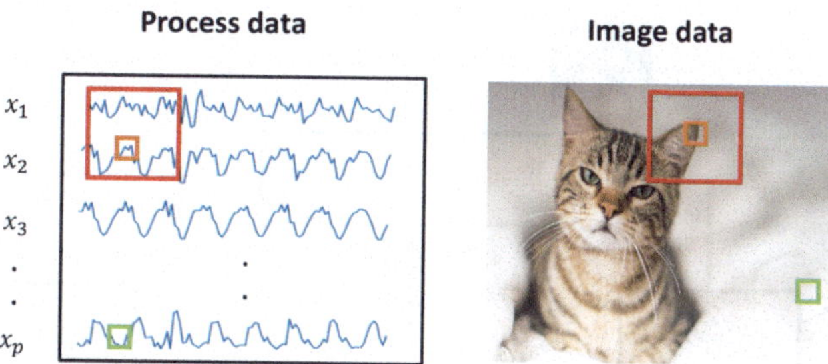

Fig. 1.7 Main differences between images and process data. (1) The simplest information unit (orange box) in images is a pixel, while in process data, it is a process measurement. (2) In image data, pixels within the same local receptive field (red box) are strongly correlated, while non-local pixels (orange and green boxes) are unlikely to be correlated. On the other hand, process data do not possess this locality property; for example, x_p can be strongly correlated with x_1 whereas x_1 and x_2 can be uncorrelated. (3) Process data contain temporal and spatial information, while image data contain spatial information only

(RGB) intensities [35]. In contrast, each observation in process data corresponds to a physical process measurement. Moreover, unlike image data, process data exhibit temporal dependencies, adding another layer of complexity. Figure 1.7 presents the key differences between image data and process data, which highlights why adopting computer vision models to industrial process monitoring requires addressing these fundamental differences.

In this monograph, we explore this analogy in depth and propose frameworks that address the representational differences between image data and process data. Specifically, we investigate how methods originally designed for image analysis can be extended to capture the temporal and spatial characteristics of process data, with the aim of improving interpretability and performance for FDD tasks.

1.3.2 Definition

Visual analytics, in the context of process monitoring, refers to a paradigm that combines visual data representations, image-based models, and domain expert reasoning to improve FDD. This involves transforming raw process data into visual representations, which are then analyzed using image-based algorithms. Visual analytics is motivated by the role of visual perception in human reasoning; encoding process data as visual clues allows process experts to examine the data in a more intuitive and informative manner.

The core concept of visual analytics frameworks is to encode process data into image-like formats prior to analysis. Rather than applying image-based tools (e.g., CNN) directly to raw time series, the data are transformed into visual patterns (i.e., pixel-based grids) that capture key features not readily evident in the time domain. This transformation leverages the pattern-recognition capabilities of image-based models while accounting for the differences between process and image data. Figure 1.8 illustrates this idea: historical

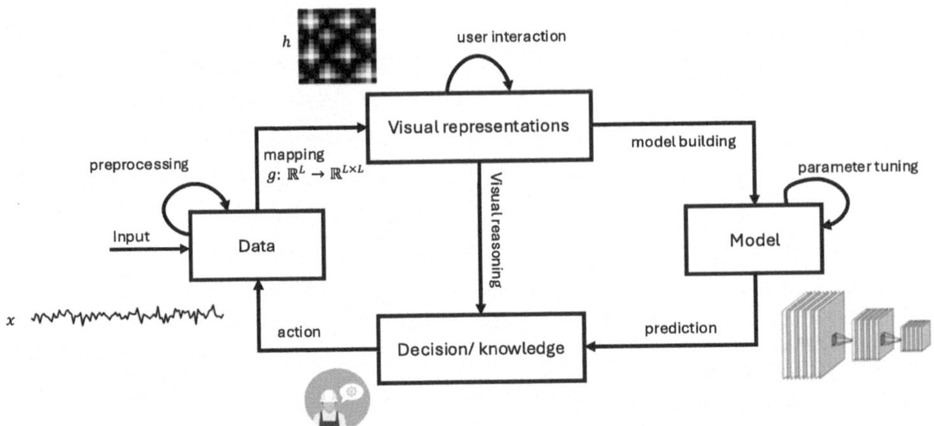

Fig. 1.8 A conceptual overview of a visual analytics framework for FDD. Given a raw process data input $x \in \mathbb{R}^L$, a transformation function $g : \mathbb{R}^L \to \mathbb{R}^{L\times L}$ encodes the signal into an image-like representation that preserves temporal dependencies and reveals discriminative patterns. The encoded data are then processed using image-based models to perform feature extraction and classification. Visual representations and model outputs are then presented for expert interaction, facilitating model interpretation, result validation, and decision support in industrial process monitoring

time series data are converted into image-like representations whose textures and patterns reflect process dynamics. These representations are then used as inputs to image-based models for FDD tasks.

A key advantage of visual analytics is improved interpretability. Image-like formats enable process experts to inspect and reason about process behaviour more intuitively than raw time series plots. Specifically, process experts can associate visual patterns, such as distinct lines, clusters, or textures, with specific operating conditions or faults. For example, line or grid patterns may suggest periodic oscillations, while scattered textures may indicate noise or irregular fault conditions. These patterns can be linked to known process behaviours, which facilitates validation of model outputs.

The integration of expert reasoning with image-based analysis not only improves explainability but can also improve prediction accuracy. Image-based models are designed to process spatially structured inputs (i.e., image data) and using image-like representations of process data as inputs ensures compatibility with their architectural design. This alignment can lead to improved model performance compared to using raw process data directly as input. Next, the model predictions can be supported by qualitative evidence (i.e., patterns in the visual representation) that aligns with engineering intuition, which leads to more informed decision-making for FDD tasks.

1.4 Book Outline

The rest of the monograph is structured into five chapters. Figure 1.9 provides a visual overview of the subsequent chapters and their thematic organization. The outline is organized to first introduce the benchmark datasets used in this monograph to test and validate the proposed methods, followed by a comprehensive discussion of three visual analytics pathways applied to FDD: feature engineering, architecture engineering, and data engineering.

Chapter 2 introduces the benchmarks used throughout this monograph. It begins by describing the simulated benchmark dataset, including how it was generated and the specific faults studied. Next, a real-world industrial benchmark derived from a pyrometallurgical process is presented. To provide context, important background information on the industrial process and associated faults is discussed. Chapters 3 to 5 constitute the core contributions of the monograph, each representing a distinct pathway within the visual analytics framework. These pathways, feature engineering, architecture engineering, and data engineering, represent different strategies for transforming time-series data into image-like representations for FDD tasks.

Chapter 3 focuses on the *feature engineering* category of visual analytics, specifically on imaging time series tools. This chapter introduces two prominent methods for transforming time series data into visual representations. Each method is discussed in detail with a review of the relevant literature to contextualize its development and applications. The chapter evaluates both methods on the benchmark datasets to analyze their performance and discusses their respective advantages and limitations. Furthermore, the visual representations generated by these methods are examined to understand the conceptual properties they capture from time series data.

Chapter 4 presents the *architecture engineering* pathway, which involves designing end-to-end deep learning models that learn to generate visual representations as part of their internal processing. This approach uses supervised learning frameworks to develop neural network architectures that extract discriminative visual patterns directly from temporal process data, without relying on fixed transformations.

Chapter 5 explores the *data engineering* pathway, which emphasizes learning visual representations through data-centric strategies, particularly in scenarios with limited

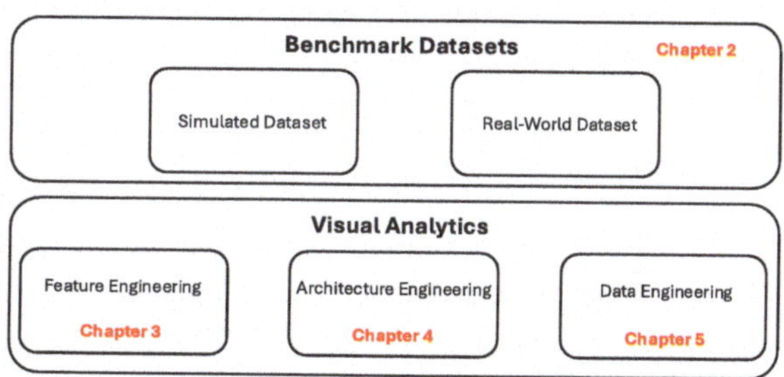

Fig. 1.9 Outline of the monograph. Chapter 2 introduces benchmark datasets. Chapters 3 to 5 explore three distinct visual analytics pathways: feature engineering, architecture engineering, and data engineering

labelled data. A novel contrastive learning framework is proposed, designed to extract informative latent representations by exploiting cross-domain relationships between temporal and frequency components of process data. This approach facilitates robust feature learning from historical data and addresses the challenges of data scarcity by leveraging self-supervised learning techniques. The experimental analysis evaluates the framework in scenarios with limited labelled data and examines the generalizability of the learned features.

Chapter 6 concludes the monograph by summarizing the main contributions and exploring potential future research directions. The key findings of this work are reviewed, and the practical applications of visual analytics in industrial settings are discussed.

References

1. Reis MS, Gins G. Industrial Process Monitoring in the Big Data/Industry 4.0 Era: from Detection, to Diagnosis, to Prognosis. Processes. 2017; 5(3):35. https://doi.org/10.3390/pr5030035
2. Gantz J, Reinsel D. The Digital Universe in 2020: Big Data, Bigger Digital Shadows, and Biggest Growth in the Far East. IDC; 2012. https://www.cs.princeton.edu/courses/archive/spring13/cos598C/idc-the-digital-universe-in-2020.pdf
3. Herweck P, Johnston L. How industrial data can help unleash productivity, innovation and sustainability. World Economic Forum. Published May 17, 2022. https://www.weforum.org/stories/2022/05/industrial-data-can-unlock-our-sustainable-future-heres-how/
4. K. Zhou, Taigang Liu and Lifeng Zhou, "Industry 4.0: Towards future industrial opportunities and challenges," 2015 12th International Conference on Fuzzy Systems and Knowledge Discovery (FSKD), Zhangjiajie, China, 2015, pp. 2147–2152, doi: 10.1109/FSKD.2015.7382284.
5. Russell E, Chiang L, Braatz R. *Data-Driven Methods for Fault Detection and Diagnosis in Chemical Processes*. New York: Springer; 2000. https://doi.org/10.1007/978-1-4471-0409-4
6. Miljković D. Fault detection methods: A literature survey. In: *2011 Proceedings of the 34th International Convention MIPRO*; 2011:750–755.
7. Seifried P. Fault Detection and Diagnosis in Chemical and Petrochemical Processes, Bd. 8 der Serie "Chemical Engineering Monographs". Reviewed by D. M. Himmelblau, edited by S. W. Churchill. Amsterdam–New York: Elsevier Scientific Publishing Company; 1978. *Chemie Ingenieur Technik*. 1979;51(7):766. https://doi.org/10.1002/cite.330510726
8. Raich A, Çinar A. Statistical process monitoring and disturbance diagnosis in multivariable continuous processes. *AIChE Journal*. 1996;42(4):995–1009. https://doi.org/10.1002/aic.690420412
9. Venkatasubramanian V, Rengaswamy R, Yin K, Kavuri SN. A review of process fault detection and diagnosis: Part I—quantitative model-based methods. *Computers & Chemical Engineering*. 2003;27(3):293–311. https://doi.org/10.1016/S0098-1354(02)00160-6
10. Yiakopoulos C, Koutsoudaki M, Gryllias K, Antoniadis I. Improving the performance of univariate control charts for abnormal detection and classification. *Mechanical Systems and Signal Processing*. 2017;86:122–150. https://doi.org/10.1016/j.ymssp.2016.09.036
11. MacGregor JF, Kourti T. Statistical process control of multivariate processes. *Control Engineering Practice*. 1995;3(3):403–414. https://doi.org/10.1016/0967-0661(95)00014-L
12. Yin S, Ding SX, Xie X, Luo H. A review on basic data-driven approaches for industrial process monitoring. *IEEE Transactions on Industrial Electronics*. 2014;61(11):6418–6428. https://doi.org/10.1109/TIE.2014.2301773

13. Qiu S, Cui X, Ping Z, Shan N, Li Z, Bao X, Xu X. Deep learning techniques in intelligent fault diagnosis and prognosis for industrial systems: A review. *Sensors*. 2023;23(3):1305. https://doi.org/10.3390/s23031305
14. Yu J, Zhang Y. Challenges and opportunities of deep learning-based process fault detection and diagnosis: a review. *Neural Computing and Applications*. 2023;35(1):211–252. https://doi.org/10.1007/s00521-022-08017-3
15. Yang R, Huang M, Lu Q, Zhong M. Rotating machinery fault diagnosis using long-short-term memory recurrent neural network. *IFAC-PapersOnLine*. 2018;51(24):228–232. https://doi.org/10.1016/j.ifacol.2018.09.582
16. Lee KB, Cheon S, Kim CO. A convolutional neural network for fault classification and diagnosis in semiconductor manufacturing processes. *IEEE Transactions on Semiconductor Manufacturing*. 2017;30(2):135–142. https://doi.org/10.1109/TSM.2017.2676245
17. Murdoch WJ, Singh C, Kumbier K, Abbasi-Asl R, Yu B. Definitions, methods, and applications in interpretable machine learning. *Proceedings of the National Academy of Sciences*. 2019;116(44):22071–22080. https://doi.org/10.1073/pnas.1900654116
18. Rudin C, Chen C, Chen Z, Huang H, Semenova L, Zhong C. Interpretable machine learning: Fundamental principles and 10 grand challenges. 2021. arXiv preprint arXiv:2103.11251. https://arxiv.org/abs/2103.11251
19. Lepore A, Palumbo B, Poggi JM, eds. *Interpretability for Industry 4.0: Statistical and Machine Learning Approaches*. Cham: Springer International Publishing; 2022.
20. Ramírez-Sanz J, Maestro-Prieto JA, Arnaiz-González Á, Bustillo A. Semi-supervised learning for industrial fault detection and diagnosis: A systemic review. *ISA Transactions*. 2023; (Review). https://doi.org/10.1016/j.isatra.2023.09.027
21. Ji C, Sun W. A review on data-driven process monitoring methods: Characterization and mining of industrial data. *Processes*. 2022;10(2):335. https://doi.org/10.3390/pr10020335
22. Huang J, Wen J, Yoon H, Pradhan O, O'Neill Z, Candan KS. Real vs. simulated: Questions on the capability of simulated datasets for building fault detection from a data-driven perspective. *Building and Environment*. 2022;206:107384. https://doi.org/10.1016/j.buildenv.2021.107384
23. Calabrese F, Regattieri A, Bortolini M, Galizia FG. Data-driven fault detection and diagnosis: Challenges and opportunities in real-world scenarios. *Applied Sciences*. 2022;12(18):Article 9212. https://doi.org/10.3390/app12189212
24. Qian J, Song Z, Yao Y, Zhu Z, Zhang X. A review on autoencoder based representation learning for fault detection and diagnosis in industrial processes. *Chemometrics and Intelligent Laboratory Systems*. 2022;231:104711. https://doi.org/10.1016/j.chemolab.2022.104711
25. Gajjar S, Kulahci M, Palazoglu A. Real-time fault detection and diagnosis using sparse principal component analysis. *Journal of Process Control*. 2018;67:112–128. https://doi.org/10.1016/j.jprocont.2017.03.005
26. Faouzi J. Time series classification: A review of algorithms and implementations. In: Rocha J, Viana CM, Oliveira S, editors. *Time Series Analysis*. Rijeka: IntechOpen; 2024. Chapter 2. https://doi.org/10.5772/intechopen.1004810
27. Middlehurst M, Large J, Flynn M, Lines J, Bostrom A, Bagnall A. HIVE-COTE 2.0: A new meta ensemble for time series classification. *Machine Learning*. 2021;110(11–12):3211–3243. https://doi.org/10.1007/s10994-021-06057-9
28. Park YJ, Fan SKS, Hsu CY. A review on fault detection and process diagnostics in industrial processes. *Processes*. 2020;8(9):Article 1123. https://doi.org/10.3390/pr8091123
29. Ismail Fawaz H, Lucas B, Forestier G, Pelletier C, Schmidt D, Weber J, Webb GI, Idoumghar L, Muller PA, Petitjean F. InceptionTime: Finding AlexNet for time series classification. *Data Mining and Knowledge Discovery*. 2020;34:1–27. https://doi.org/10.1007/s10618-020-00710-y

30. Dempster A, Petitjean F, Webb GI. ROCKET: Exceptionally fast and accurate time series classification using random convolutional kernels. *Data Mining and Knowledge Discovery*. 2020;34(5):1454–1495. https://doi.org/10.1007/s10618-020-00701-z
31. Jiang S, Zavala VM. Convolutional neural nets in chemical engineering: Foundations, computations, and applications. *AIChE Journal*. 2021;67(9):e17282. https://doi.org/10.1002/aic.17282
32. Zheng Y, Liu Q, Chen E, Ge Y, Zhao JL. Time series classification using multi-channels deep convolutional neural networks. *Lecture Notes in Computer Science (WAIM)*. 2014;8485:298–310.
33. Yuan X, Qi S, Shardt YAW, Wang Y, Yang C, Gui W. Soft sensor model for dynamic processes based on multichannel convolutional neural network. *Chemometrics and Intelligent Laboratory Systems*. 2020;203:104050. https://doi.org/10.1016/j.chemolab.2020.104050
34. Wang K, Shang C, Liu L, Jiang Y, Huang D, Yang F. Dynamic soft sensor development based on convolutional neural networks. *Industrial & Engineering Chemistry Research*. 2019;58(26):11521–11531. https://doi.org/10.1021/acs.iecr.9b02513
35. Forsyth DA, Ponce J. *Computer Vision: A Modern Approach*, 2nd ed. Pitman; 2012.

Evaluation Benchmarks

2

Contents

2.1 Benchmark Datasets for Process Monitoring

Benchmark datasets have played a central role in advancing data-driven research communities by enabling systematic model development and comparison. A prominent example is the ImageNet dataset, released in 2010, which rapidly became a standard benchmark for image classification research [1]. Over the following eleven years, the accuracy of top-performing models improved significantly, increasing from 50.90% to 90.88% [2]. This progress was facilitated, in part, by the widespread adoption of a standardized, open-source dataset that supported reproducible evaluation. Over the same period, process industries have undergone substantial digital transformation, which led to a significant increase in the availability of data for FDD.

Existing FDD benchmark datasets can be broadly categorized according to their data acquisition methodology. One category consists of simulated datasets, which are artificially generated to resemble real processes while preserving known fault patterns [3]. These datasets span a wide range of fidelity levels depending on how accurately the simulation models represent real-world behaviour [4]. Low-fidelity benchmarks, such as the Tennessee Eastman Process (TEP) [5, 6] and PenSim [7, 8], are typically derived from first-principles models that simplify uncertainty and disturbances using statistical or piecewise-linear approximations. Higher-fidelity simulations incorporate detailed physical descriptions of equipment and instrumentation, grounded in principles such as thermodynamics and mechanics. The DAMADICS benchmark, for example, models

I. Yousef et al., *Visual Analytics for Process Monitoring*, Synthesis Lectures on Emerging Engineering Technologies, https://doi.org/10.1007/978-3-032-22126-1_2

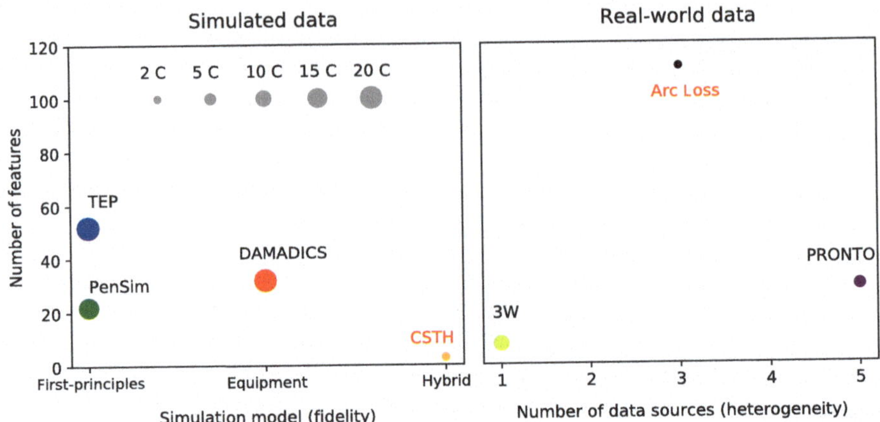

Fig. 2.1 A conceptual comparison of FDD benchmarks reported in the literature. The benchmarks are compared based on the following criteria: (1) data origin (simulated data versus measurements from physical sensors), (2) number of features, (3) number of classes, (4) simulation model type for simulated benchmarks, and (5) data heterogeneity for real-world benchmarks. The size of each marker reflects the number of classes, with the legend shown in the upper left corner ranging from 2 to 20 classes. Both panels share the same y-axis

electro-pneumatic actuators operating within a sugar production process [9]. Hybrid (high-fidelity) benchmarks further enhance realism by combining simulation models with measurements from real processes. The continuous stirred tank heater (CSTH) benchmark [10], a hybrid benchmark, provides a platform for evaluating FDD methods under realistic noise levels, disturbances, and operational constraints [11].

A second category of benchmarks is based on real-world industrial data collected directly from operating processes. Such datasets may include heterogeneous information from multiple sources, including sensor measurements, alarms, laboratory results, valve positions, images, and video recordings. For instance, the PRONTO benchmark [12] integrates data from several sources within an industrial-scale multiphase flow facility, whereas the 3W benchmark [13] focuses on a small set of measurements associated with offshore oil well operations. Data heterogeneity increases the complexity of analysis, thereby reflecting the challenges encountered in practical FDD applications. A conceptual comparison of several widely used process monitoring benchmarks is shown in Fig. 2.1. Readers seeking a comprehensive overview of publicly available benchmarks are referred to the review by Melo et al. [14].

Despite their practical relevance, real-world benchmarks remain less common in the process systems engineering (PSE) literature than simulated datasets. This can be attributed to the difficulty of acquiring real-world industrial data that contain frequent and clearly labeled fault events [15]. In practice, industrial systems often exhibit long mean times between failures [16], faults may result in complete process shutdowns, and data owners are frequently unwilling to release operational data due to confidentiality and intellectual property concerns [14]. As a result, many benchmark datasets rely on faults that are synthetically introduced through parameter perturbations within simulation environments.

In this monograph, we validate our proposed models on two types of benchmark datasets: a simulated dataset and a real-world dataset. Specifically, we employ the CSTH process as a simulated benchmark dataset. Furthermore, we introduce the Arc Loss benchmark, a novel real-world dataset obtained from three heterogeneous data sources: process measurements, valve positions, and laboratory results. The Arc Loss benchmark dataset was first introduced in [17]. This dataset provides a realistic testing platform for the proposed models under industrial conditions. Further details on both benchmarks (CSTH and Arc Loss) and associated fault scenarios are provided in the following sections.

2.2 A Simulated Benchmark: The CSTH Dataset

The CSTH system is a nonlinear dynamic process described in [18]. Figure 2.2 show the feedback control configuration of the CSTH system. In this setup, hot water (HW) and cold water (CW) streams are mixed, heated by steam passing through a coil, and then discharged from the tank. To maintain stability, a closed-loop control scheme regulates the tank temperature, liquid level, and CW flow. The manipulated inputs are the steam, HW, and CW valve openings, while the measured outputs (controlled variables) include the CW flow, tank level, and temperature. The CSTH model is considered a semi-empirical (hybrid) representation, combining fundamental first-principles equations with empirical relationships derived from experimental data.

The left-hand column of Fig. 2.3 shows the tank level, temperature, and CW flow measurements obtained under normal operating conditions. To construct a binary fault

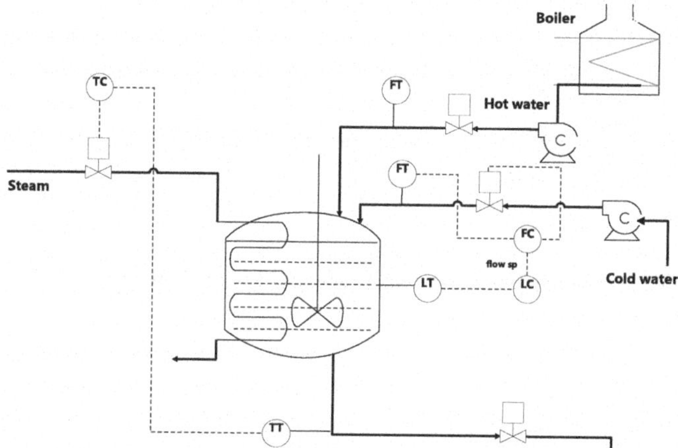

Fig. 2.2 The CSTH feedback control system

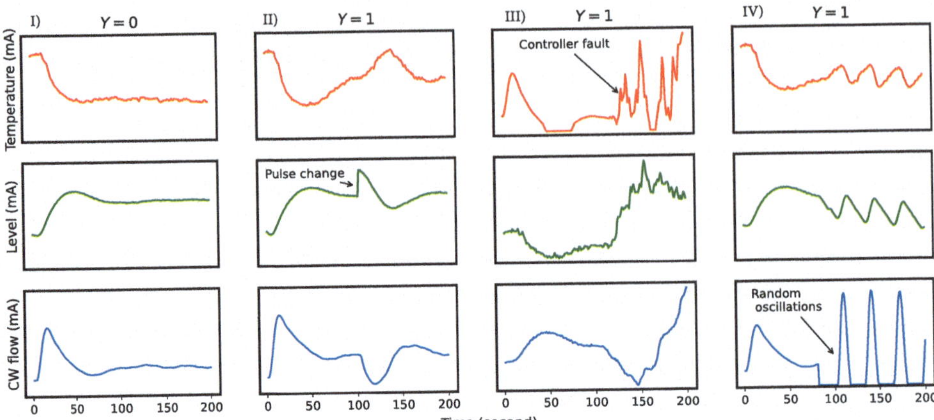

Fig. 2.3 A visual comparison of the CSTH system under smooth operation ($Y = 0$, left-hand column) and faulty operation ($Y = 1$, second to fourth columns)

Table 2.1 The CSTH dataset summary

Total number of samples (N)	9000
Training: validation: testing ratio	70:10:20
Number of variables (p)	3
Sampling period	1 sec
Signal length (L)	200
Number of classes (C)	2
Class ratio	50:50 (balanced)

classification problem, three fault scenarios are considered: (1) an abrupt pulse disturbance added to the level transmitter signals (Fig. 2.3II), (2) a temperature controller malfunction represented by random changes in its parameters (Fig. 2.3III), and (3) random sinusoidal noise injected into the CW flow controller output signals (Fig. 2.3IV). A summary of the dataset is provided in Table 2.1. This benchmark problem is formulated as a binary classification task, where normal operating data are labelled as $Y = 0$ and faulty operating data are labelled as $Y = 1$.

The CSTH dataset is publicly available and can be accessed and downloaded at https://doi.org/10.5683/SP3/8FXNGM.

2.3 An Industrial Benchmark: The Arc Loss Dataset

The Arc Loss benchmark dataset originates from a large-scale open-pit mining and pyrometallurgical plant. Figure 2.4 provides an overview of the mining and smelting operations. In these processes, high-grade oxidized ore deposits are converted into refined base metals, which are subsequently processed to shotting and packaging units before being shipped to end-users [19].

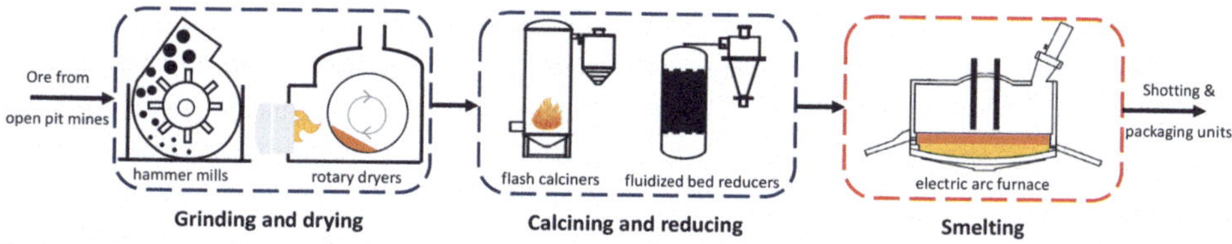

Fig. 2.4 An illustration of the broader mining and metallurgical processes

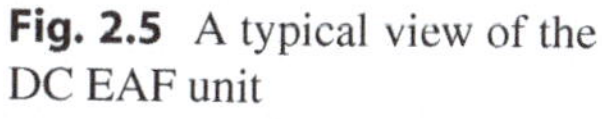

Fig. 2.5 A typical view of the DC EAF unit

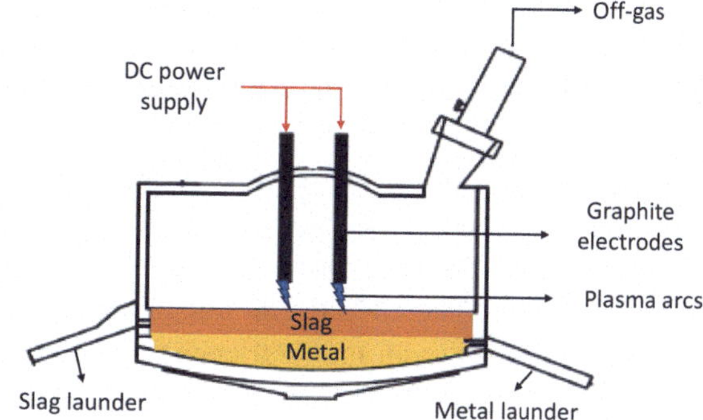

Extraction begins with saprolite ore mined from several open pits using hydraulic shovels. The ore is then hauled by dump trucks to the ore preparation plant. Due to the friable nature of the run-of-mine ore, waste rock and undersized particles are removed via crushing and screening operations [20]. The crushed ore is then conveyed on an overland belt conveyor to the metallurgical plant for further processing.

In the metallurgical process, the ore undergoes a series of sequential processing operations, including milling, drying, calcining, reduction, and smelting. Since the ore from the preparation stage can contain up to 40% moisture, drying is performed first [21]. Hammer mill flash dryers reduce the free moisture content to below 1%. The dried ore is then fed into a series of calciner cyclones operating at approximately 1000 °C, where it is dehydrated [22]. Following calcination, pre-reduction takes place in fluidized bed reactors using pulverized coal and hot gases at around 1000 °C to remove oxides [23]. This step is essential to ensure operational stability of the subsequent smelting operation in the direct current electric arc furnace (DC EAF). The focus of this case study is on the operation of the DC EAF unit, which is used as a smelter to refine ores into base metals.

A schematic of the DC EAF is shown in Fig. 2.5. The DC EAF unit is a refractory-lined cylindrical vessel with water-cooled sidewalls, a conical roof, and twin hollow graphite electrodes located vertically through the roof [24]. Inside the steel vessel, a molten bath forms, consisting of a denser metal phase below a lighter slag phase. Refined ore is fed into the furnace through roof-mounted feed ports equipped with weight-bin feeders, while slag and metal are tapped intermittently through dedicated launders. Two plasma arcs span from the lower tips of the electrodes to the molten bath surface, serving as cathode and anode, respectively. The electrodes are connected to a large DC power supply that provides the electrical energy required to sustain the furnace operation [25].

The plasma arcs are developed by direct current transfer from the cathodes to the anode and serve as the main heating component in the DC EAF unit [24]. The arcs convert electrical power from the DC supply (up to 80 MW) into thermal energy exceeding 1500 °C, which is required to maintain the desired metal-to-slag ratio in the furnace [26]. Since the operation of the DC EAF is directly related to the thermal energy transferred into the molten bath, the presence of the plasma arcs is critical to ensure operational stability and maximum production efficiency.

2.3.1 The Process Fault: An Unexpected Loss of Plasma Arcs

The plasma arcs, in the DC EAF unit, are high-temperature, high-velocity plasma jets that efficiently conduct electricity [27]. They serve as the main heating source, converting electrical power from the DC power supply into thermal energy and transferring it into the molten bath. The DC electric circuit through the furnace connects the arc and the slag bath in series, with the total operating voltage divided between them according to the electrical properties, primarily resistivity, of the slag [24]. The slag is often of high electrical resistivity, resulting in a significant voltage drop across the slag layer of the circuit [28]. This can reduce the available voltage to the arcs below the threshold needed for the arcs to develop, resulting in an arc loss event. Additional suspected causes of arc loss include upstream process disturbances, electrical instabilities, and excessively long arcs.

Direct observation of the plasma arcs is not possible due to the extreme operating conditions and the safety-critical nature of the process. Therefore, the daily raw data exports do not include explicit arc loss labels. Instead, arc loss events are inferred from power fluctuations. Specifically, a loss of arc in an electrode at time t is defined by three power-related conditions: (1) a sudden power drop of 10 MW or more compared to the value at $t - 0.6$ min, (2) stable power for about 11.25 minutes within a standard deviation of 2 MW before the drop (i.e., between $t - 11.85$ min and $t - 0.6$ min), and (3) recovery of power to within $\pm$ 5 MW of the original stable range within approximately 9.85 minutes after the drop ($t + 9.25$ min) [29]. These conditions are illustrated in Fig. 2.6. Based on this definition, each sample is assigned a binary label, where $Y = 0$ denotes normal operation and $Y = 1$ indicates the occurrence of an arc loss.

The objective is to use a full year of high-frequency operating data to predict the onset of an arc loss event, represented by the target variable Y. An arc loss event is defined as an event where one or more of the following conditions is met: (1) loss of the arc in electrode A only, (2) loss of the arc in electrode B only, or (3) simultaneous arc losses in both electrodes A and B. The target variable Y is formally defined as

$$Y = \begin{cases} 0, & Y^A + Y^B = 0 \\ 1, & Y^A + Y^B > 0 \end{cases} \tag{2.1}$$

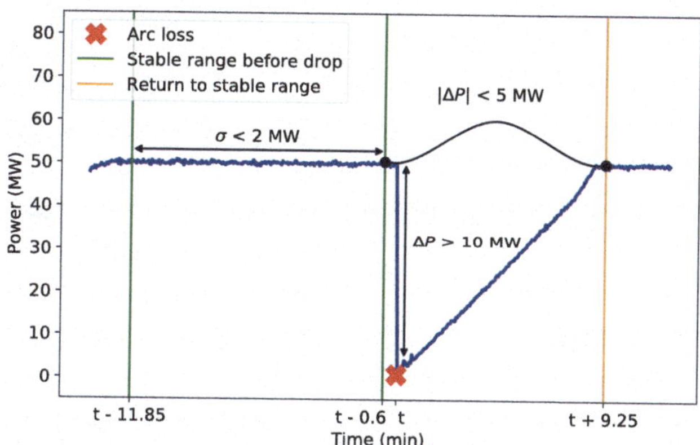

Fig. 2.6 Timeline chart of the power-based criteria used to identify arc loss events

where Y^A and $Y^B \in [0, 1]$ are binary indicators corresponding to arc loss events in electrodes A and B, respectively.

The unforeseen occurrence of an arc loss event has substantial consequences on the operation of the DC EAF. Beyond reducing electrical efficiency, recovery from an arc loss typically requires a temporary decrease in furnace feed rate accompanied by an increase in power input. Figure 2.7 illustrates the operational consequences of arc loss by comparing process measurements recorded under stable and faulty operating conditions. Given that furnace operation is directly dependent on the presence of sustained plasma arcs, the development of a reliable predictive alarm capable of identifying the onset of arc loss would yield considerable economic and environmental benefits.

2.3.2 Exploratory Data Analysis

The raw Arc Loss dataset consists of one year of high-frequency operating data collected from various metallurgical process units (e.g., calcining, smelting, reduction, etc.). This section provides an overview of the statistics and characteristics of the raw data.

2.3.2.1 Raw Data Structure

Each daily export file captures one full day of operation and contains over 200 columns and approximately 30,000 rows. The columns in the daily exports correspond to process variables and their associated timestamps. As shown in Fig. 2.8, the sampling rates of different variables vary. The number of rows (represented by the height of the blue and red bars) indicates the number of samples of a given variable collected during the day. Some variables are recorded at high frequency (e.g., every three seconds), whereas laboratory measurements may have sampling intervals longer than three hours. Overall, the total daily exports have an uncompressed size of 17.4 gigabytes (GB).

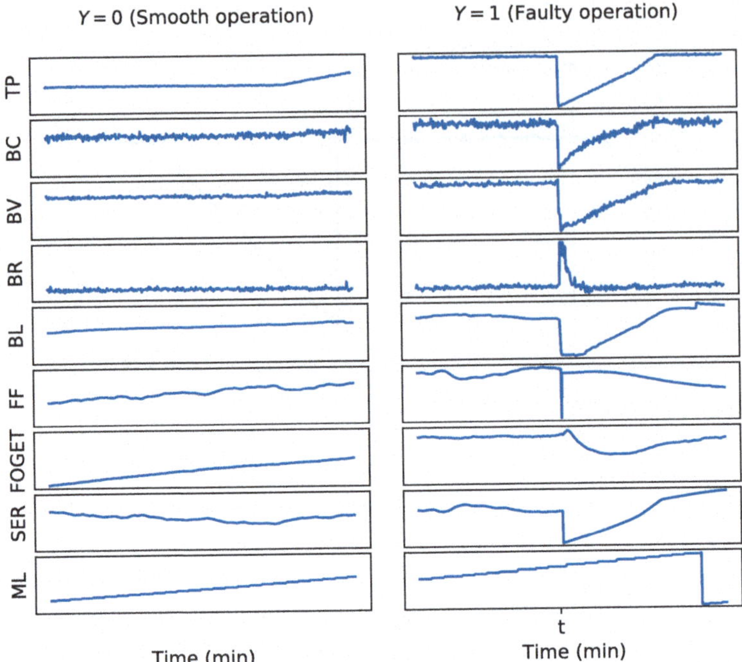

Fig. 2.7 Comparison of process measurements during stable operation (left) and during an arc loss event (right). The time instant t denotes the occurrence of the arc loss. Both plots share a common y-axis. Variable descriptions are provided in Table A1

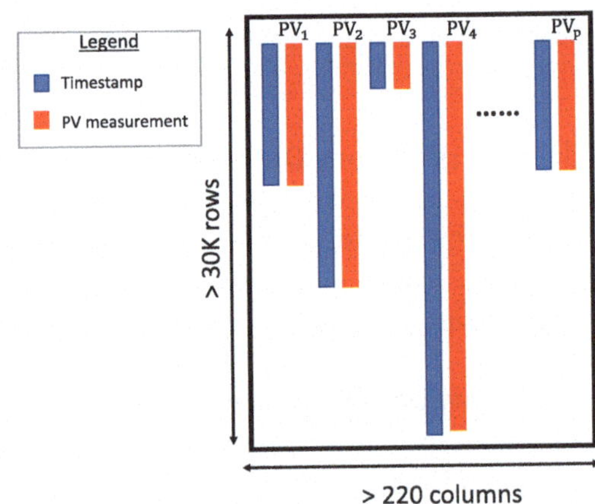

Fig. 2.8 An illustration of raw daily export structure for the Arc Loss dataset

The raw dataset consists of a time index and 111 process variables. Table 2.2 presents key statistics for the raw Arc Loss benchmark dataset. For brevity, readers are referred to Table A1 for detailed information on each process variable, including its description, range, and unit of measurement. Among the 111 process variables, 92 are high-frequency variables that capture physical process characteristics (e.g., feed rate, temperature),

Table 2.2 Overview of the raw Arc Loss dataset statistics

Total number of samples (N)	10,483,200
Total number of process variables (p)	111
Total number of categorical variables	14
Sampling period	3 sec
Training/validation set ratio	83.5% (Jan.–Oct.)
Testing set ratio	16.5% (Nov.–Dec.)
Number of classes (C)	2
Class ratio ($Y = 0$:$Y = 1$)	3337:1

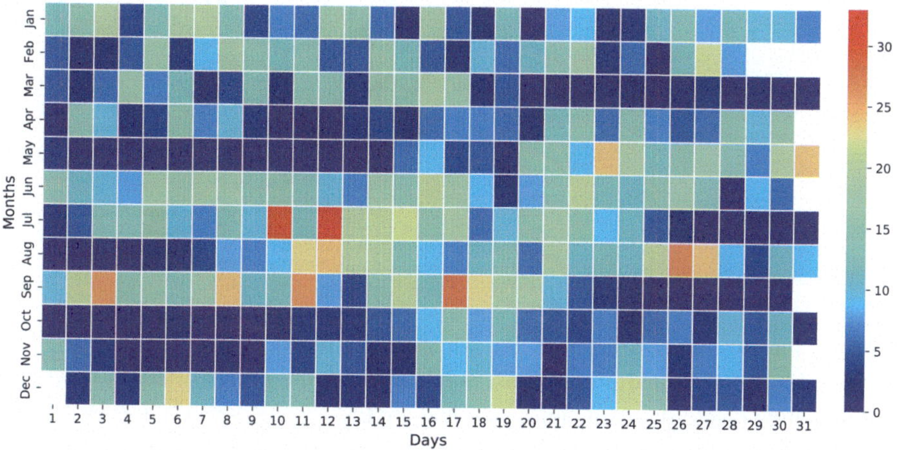

Fig. 2.9 The number of arc loss events per day throughout a year of operation

14 are categorical variables corresponding to valve positions, and five are laboratory measurements with sampling intervals greater than one hour.

The raw Arc Loss benchmark dataset is publicly available and can be downloaded at https://codalab.lisn.upsaclay.fr/competitions/13129.

2.3.2.2 Data Visualization

Figure 2.9 shows the daily occurrences of arc loss events over an entire year of operation. Arc loss faults are observed most frequently between May and September. Next, Fig. 2.10 presents a circular plot showing the arc loss rate for June. The chart represents a series of 30 days, beginning on June 1st, 2022 at 00:00 in the innermost circle and moving clockwise to June 30th, 2022 at 23:00 in the outermost ring. Darker blue regions denote periods with higher fault frequency, whereas lighter shades correspond to lower fault occurrence. Overall, Figs. 2.9 and 2.10 reveal that arc loss faults exhibit randomness, seasonality, and variability, making fault detection a non-trivial task.

Over one year of operation, the DC EAF unit recorded a total of 3141 arc loss events. As shown in Fig. 2.11a, June experienced the highest number of faulty events, followed by September and July, while October recorded the fewest. Figure 2.11b shows that both

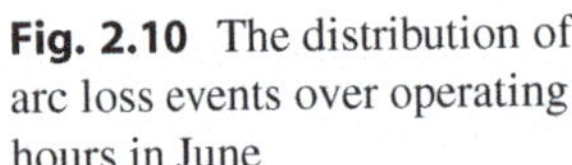

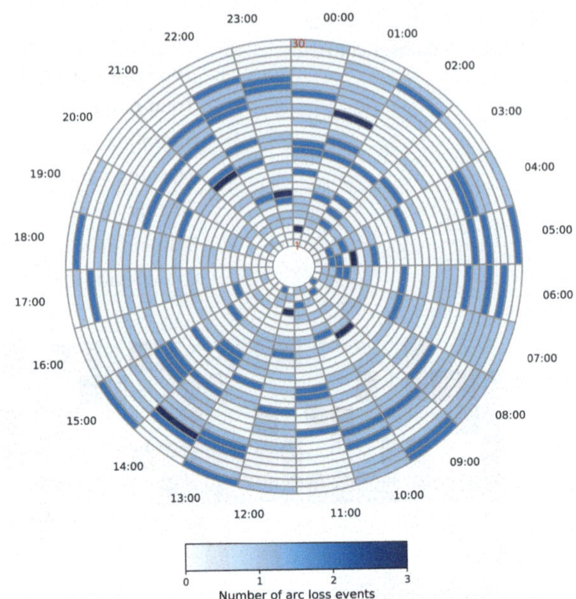

Fig. 2.10 The distribution of arc loss events over operating hours in June

subclasses (arc loss A and arc loss B), which make up the faulty class ($Y = 1$), are represented in nearly equal proportions, indicating that the process fault is independent of the operating electrode. Although individual arc loss events typically last less than one minute, each fault can result in up to ten minutes of lost production time. On average, 9.4 arc loss events were recorded per day, corresponding to approximately 82 minutes of lost production, as shown in Fig. 2.11c and d.

2.3.2.3 Data Preprocessing

The raw Arc Loss dataset is riddled with problematic artifacts. In all industrial machine learning applications, it is necessary to apply at least one data preprocessing technique to ensure data usability [30]. The main objective of data preprocessing is to transform large amounts of raw, messy, and noisy process data into a form that is suitable for statistical machine learning algorithms. Although onerous, it is an essential step that improves model performance by removing inadequate, irrelevant, redundant, or confounding data. In what follows, we outline the preprocessing steps performed on the raw Arc Loss dataset, which include data structuring and imputation, cleaning, feature scaling, and segmentation.

As illustrated in Fig. 2.12, each process variable in the raw data is associated with a timestamp, but the sampling rates vary widely among process variables. To provide structure while reducing redundancy, the most densely sampled variable is identified for each day, and its timestamp is adopted as a unified reference for all other variables (blue bar on the right-hand side of Fig. 2.12). Less densely sampled variables are resampled using a forward-fill (zero-order hold) method to align them with the unified time index, resulting in a structured data-frame matrix with equal row counts and a single timestamp

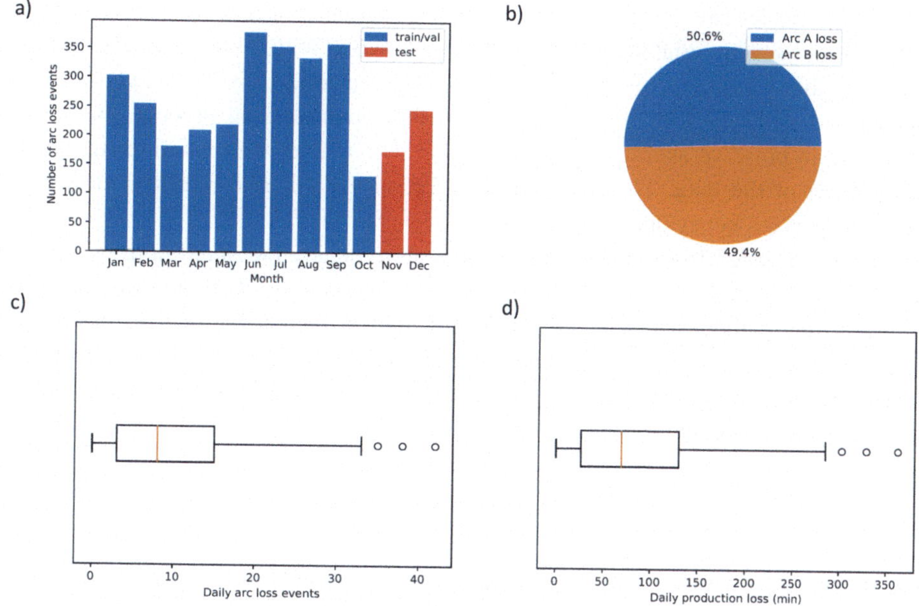

Fig. 2.11 Statistical summary of the Arc Loss dataset

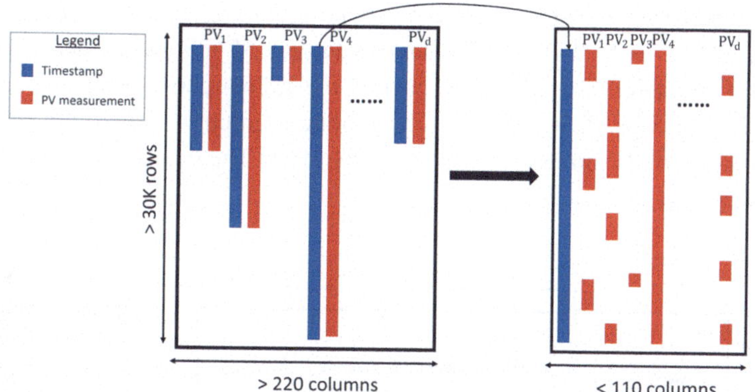

Fig. 2.12 An illustration of structuring a raw daily export

column. Inconsistent entries (e.g., "tag not found," "bad quality") are replaced with NaN values and then filled using the same forward-fill strategy, while categorical variables such as "open" and "close" are converted to numerical indicators (1 and 0, respectively).

The data cleaning step ensures that the processed data are both consistent and reliable. Process variables limits (presented in Table A1) are used to filter out nonsensical values (e.g., negative feed rates) and remove outliers. Outliers are identified using the Z-score method, where data points exceeding three standard deviations from the mean are considered outliers and replaced using zero-order hold [31]. Furthermore, irrelevant or unreliable variables are removed—five laboratory variables due to their low sampling frequency and seven variables (W1F, W2F, W3F, W6F, W7F, W8F, and TMF) due to their unreliable measurements. Shutdown periods characterized by low power levels are detected and replaced with zeros. Throughout this preprocessing process, the guiding principle is to preserve as much raw information as possible while minimizing external manipulation of the data. This is to ensure a broad yet accurate representation of the underlying process behaviour.

Variables recorded from industrial processes are typically measured in different units and with varying magnitudes, which can lead to unequal contributions to the analysis. Ultimately, these inconsistencies in scale may induce significant bias in the learned models. To address this, we apply feature scaling as a preprocessing step to standardize the variables onto a unified scale [32]. Each variable is normalized to have a zero mean and a unit standard deviation using the equation:

$$x_{\mathbf{scaled}} = \frac{x - \mu}{\sigma} \tag{2.2}$$

where x represents a process variable measurement, and μ and σ are its mean and standard deviation.

Finally, the data segmentation step addresses the severe class imbalance present in the raw dataset, where 99.67% of samples correspond to normal operation ($Y = 0$) and only 0.33% to arc loss ($Y = 1$). To create a balanced dataset, 55-minute segments are extracted from the 5–60 minute interval preceding each of the 1526 arc loss events, representing the minority class. An equivalent number of 55-minute segments is randomly selected from stable operation periods to represent the majority class. The data segmentation process is displayed in Fig. 2.13.

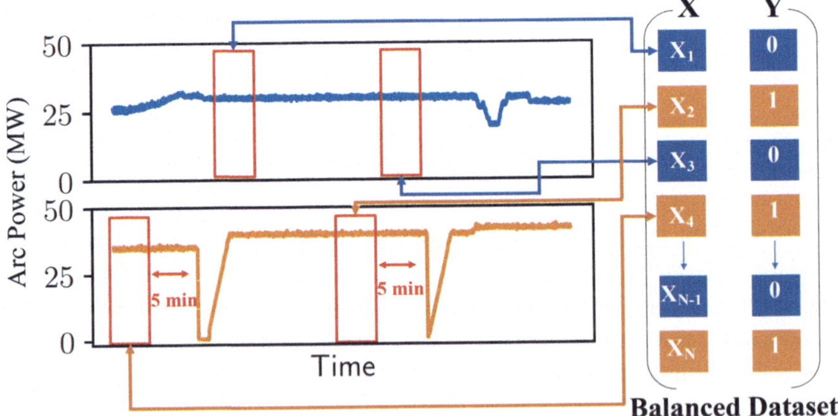

Fig. 2.13 Illustration of data segmentation to create a balanced dataset

Table 2.3 The processed Arc Loss dataset summary

Total number of samples (N)	3226
Training: validation: testing ratio	70:10:20
Number of variables (p)	96
Sampling period	3 sec
Signal length (L)	1101
Number of classes (C)	2
Class ratio	50:50 (balanced)

The processed Arc Loss dataset is a balanced MTS dataset with 3226 samples, each comprising 1101 time steps (corresponding to 55 minutes) and 96 process variables. The processed Arc Loss dataset is further divided chronologically, with 70% allocated for training, 10% for validation, and 20% for testing. A detailed summary of the processed dataset is provided in Table 2.3. The preprocessed dataset is publicly available and can be accessed at the following link: https://doi.org/10.5683/SP3/NREPZM. Please note that we use the processed Arc Loss dataset as a benchmark throughout this monograph.

References

1. Russakovsky O, Deng J, Su H, Krause J, Satheesh S, Ma S, Huang Z, Karpathy A, Khosla A, Bernstein MS, Berg AC, Fei-Fei L. ImageNet large scale visual recognition challenge. *arXiv preprint arXiv:1409.0575*. 2014. http://arxiv.org/abs/1409.0575
2. Dai Z, Liu H, Le QV, Tan M. CoAtNet: Marrying convolution and attention for all data sizes. *arXiv preprint arXiv:2106.04803*. 2021. https://arxiv.org/abs/2106.04803
3. Olson R, La Cava W, Orzechowski P, Urbanowicz R, Moore J. PMLB: A large benchmark suite for machine learning evaluation and comparison. *BioData Mining*. 2017;10:36. https://doi.org/10.1186/s13040-017-0154-4
4. Feinstein A, Cannon H. Fidelity, verifiability, and validity of simulation: Constructs for evaluation. *Developments in Business Simulation and Experiential Learning*. 2001;28:57–67.
5. Downs JJ, Vogel EF. A plant-wide industrial process control problem. *Computers & Chemical Engineering*. 1993;17(3):245–255. https://doi.org/10.1016/0098-1354(93)80018-I
6. Chiang LH, Russell EL, Braatz RD. Fault detection and diagnosis in industrial systems. *Measurement Science and Technology*. 2001;12(10):1745. https://doi.org/10.1088/0957-0233/12/10/706
7. Birol G, Ündey C, Çinar A. A modular simulation package for fed-batch fermentation: penicillin production. *Computers & Chemical Engineering*. 2002;26(11):1553–1565. https://doi.org/10.1016/S0098-1354(02)00127-8
8. Van Impe J, Gins G. An extensive reference dataset for fault detection and identification in batch processes. *Chemometrics and Intelligent Laboratory Systems*. 2015;148:20–31. https://doi.org/10.1016/j.chemolab.2015.08.019
9. Bartys M, Patton R, Syfert M, de las Heras S, Quevedo J. Introduction to the DAMADICS actuator FDI benchmark study. *Control Engineering Practice*. 2006;14(6):577–596. https://doi.org/10.1016/j.conengprac.2005.06.015

10. Thornhill NF, Patwardhan SC, Shah SL. A continuous stirred tank heater simulation model with applications. *Journal of Process Control*. 2008;18(3):347–360. https://doi.org/10.1016/j.jprocont.2007.07.006
11. von Rueden L, Mayer S, Sifa R, Bauckhage C, Garcke J. Combining machine learning and simulation to a hybrid modelling approach: Current and future directions. In: Berthold MR, Feelders A, Krempl G, editors. *Advances in Intelligent Data Analysis XVIII*. Springer; 2020. p. 548–560.
12. Stief A, Tan R, Cao Y, Ottewill JR, Thornhill NF, Baranowski J. A heterogeneous benchmark dataset for data analytics: Multiphase flow facility case study. *Journal of Process Control*. 2019;79:41–55. https://doi.org/10.1016/j.jprocont.2019.04.009
13. Vargas REV, Munaro CJ, Ciarelli PM, Medeiros AG, do Amaral BG, Barrionuevo DC, de Araújo JCD, Ribeiro JL, Magalhães LP. A realistic and public dataset with rare undesirable real events in oil wells. *Journal of Petroleum Science and Engineering*. 2019;181:106223. https://doi.org/10.1016/j.petrol.2019.106223
14. Melo A, Câmara MM, Clavijo N, Pinto JC. Open benchmarks for assessment of process monitoring and fault diagnosis techniques: A review and critical analysis. *Computers & Chemical Engineering*. 2022;165:107964. https://doi.org/10.1016/j.compchemeng.2022.107964
15. Lei Y, Li N, Guo L, Li N, Yan T, Lin J. Machinery health prognostics: A systematic review from data acquisition to RUL prediction. *Mechanical Systems and Signal Processing*. 2018;104:799–834. https://doi.org/10.1016/j.ymssp.2017.11.016
16. Jiao J, Zhao M, Lin J, Liang K. A comprehensive review on convolutional neural network in machine fault diagnosis. *Neurocomputing*. 2020;417:36–63. https://doi.org/10.1016/j.neucom.2020.07.088
17. Yousef I, Rippon LD, Prévost C, Shah SL, Gopaluni RB. The arc loss challenge: A novel industrial benchmark for process analytics and machine learning. *Journal of Process Control*. 2023;128:103023. https://doi.org/10.1016/j.jprocont.2023.103023
18. Thornhill NF, Patwardhan SC, Shah SL. A continuous stirred tank heater simulation model with applications. *Journal of Process Control*. 2008;18(3):347–360.
19. Liang X, Tang J, Li L, Wu Y, Sun Y. A review of metallurgical processes and purification techniques for recovering Mo, V, Ni, Co, Al from spent catalysts. *Journal of Cleaner Production*. 2022;376:134108. https://doi.org/10.1016/j.jclepro.2022.134108
20. Qu G, Zhou S, Wang H, Li B, Wei Y. Production of ferronickel concentrate from low-grade nickel laterite ore by non-melting reduction magnetic separation process. *Metals*. 2019;9(12):1340. https://doi.org/10.3390/met9121340
21. Kotzé IJ. Pilot plant production of ferronickel from nickel oxide ores and dusts in a DC arc furnace. *Minerals Engineering*. 2002;15(11):1017–1022. https://doi.org/10.1016/S0892-6875(02)00127-9
22. Keskinkilic E. Nickel laterite smelting processes and some examples of recent possible modifications to the conventional route. *Metals*. 2019;9(9):974. https://doi.org/10.3390/met9090974
23. Meihack WFAT. The potential role of fluidized beds in the metallurgical industry. *Journal of the Southern African Institute of Mining and Metallurgy*. 1986;86(5):153–160.
24. Reynolds QG, Hockaday CJ, Jordan DT, Barker IJ. Arc detection in DC arc furnaces. In: Mackey PJ, Grimsey EJ, Jones RT, Brooks GA, editors. *Celebrating the Megascale*. Springer; 2016. p. 157–167.
25. Jones RT. Reductive smelting for the recovery of nickel in a DC arc furnace. *European Metallurgical Conference*. 2013;:1019–1026.
26. Jones RT. DC arc furnaces — past, present, and future. In: Mackey PJ, Grimsey EJ, Jones RT, Brooks GA, editors. *Celebrating the Megascale*. Springer; 2016. p. 129–139.

27. Reynolds Q, Jones R, Reddy B. Mathematical and computational modelling of the dynamic behaviour of direct current plasma arcs. *Journal of the Southern African Institute of Mining and Metallurgy*. 2010;110:1–9.
28. Pauna H, Willms T, Aula M, Echterhof T, Huttula M, Fabritius T. Electric arc length-voltage and conductivity characteristics in a pilot-scale AC electric arc furnace. *Metallurgical and Materials Transactions B*. 2020;51(4):1646–1655. https://doi.org/10.1007/s11663-020-01859-z
29. Rippon LD, Yousef I, Hosseini B, Bouchoucha A, Beaulieu JF, Prévost C, Ruel M, Shah SL, Gopaluni RB. Representation learning and predictive classification: Application with an electric arc furnace. *Computers & Chemical Engineering*. 2021;150:107304. https://doi.org/10.1016/j.compchemeng.2021.107304
30. Famili A, Shen W-M, Weber R, Simoudis E. Data preprocessing and intelligent data analysis. *Intelligent Data Analysis*. 1997;1(1):3–23. https://doi.org/10.1016/S1088-467X(98)00007-9
31. Acuña E, Rodriguez C. On detection of outliers and their effect in supervised classification. In: *Proceedings of a Conference on Data Analysis*. 2004. https://api.semanticscholar.org/CorpusID:16780032
32. Sola J, Sevilla J. Importance of input data normalization for the application of neural networks to complex industrial problems. *IEEE Transactions on Nuclear Science*. 1997;44(3):1464–1468. https://doi.org/10.1109/23.589532

Visual Analytics Pathway I: Feature Engineering 3

Contents

3.1 Gramian Angular Field (GAF)

A GAF is a two-dimensional visual representation of a univariate time series, originally introduced by Wang and Oates, which captures information about the static behaviour of the time series [1]. Figure 3.1 illustrates the step-by-step instructions for encoding a univariate time series as a GAF representation. First, the scaled time series $\hat{x} = \{\hat{x}_1, \hat{x}_2, \ldots, \hat{x}_L\}$ is transformed from the space coordinate system to polar coordinates. The time step t_i is encoded as the radius r_i and the scaled value $\hat{x}_i$ of the time series is encoded as the angular cosine θ_i, given by:

$$\begin{aligned} r_i &= \tfrac{t_i}{L}; & i &\in L \\ \theta_i &= \cos^{-1}(\hat{x}_i); & \hat{x}_i &\in [0, 1] \end{aligned} \tag{3.1}$$

Here, L represents the length of the time series. Note that GAF is applied to scaled time series (i.e., time series values are within the interval [0, 1]) to ensure a unique result in the polar coordinate system. This bijective property (i.e., one-to-one correspondence between the scaled value and its angular representation) is due to the monotonic nature of $\cos^{-1}(x)$ when $x \in [0, 1]$. Once $\hat{x}$ is transformed into polar coordinates, the square GAF matrix is constructed. In this matrix, each entry represents the pairwise cosine distance between two angles, as given by:

I. Yousef et al., *Visual Analytics for Process Monitoring*, Synthesis Lectures on Emerging Engineering Technologies, https://doi.org/10.1007/978-3-032-22126-1_3

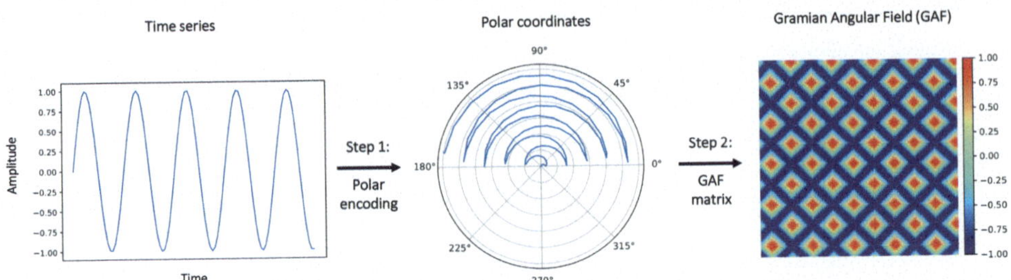

Fig. 3.1 An illustration of the transformation from a sinusoidal time series signal into a two-dimensional GAF representation. Left: a scaled 1D time series $\hat{x} = \{\hat{x}_1, \hat{x}_2, \ldots, \hat{x}_{20}\}$ with $L = 20$ time steps. Middle: the first step is to represent $\hat{x}$ in polar coordinates using Eq. 3.1. Right: the GAF representation, a square matrix of size 20×20, is obtained using Eq. 3.2. Adapted from [2], with permission

$$\text{GAF}\,[i, j] = \cos(\theta_i + \theta_j); \qquad i, j = 1, 2, \ldots, L \tag{3.2}$$

Each pixel in the GAF representation represents the sum of the directions of two time steps. GAF offers several advantages that make it a valuable tool for capturing temporal dynamics. First, GAF preserves the temporal order of the original time series, which ensures that the sequential nature of the data is maintained in the resulting visual representation. Specifically, in the GAF representation, time progresses from the top-left to the bottom-right. Additionally, GAF is invariant to monotonic transformations, allowing it to capture the same underlying patterns regardless of scaling or shifting in the time series values.

3.2 Recurrence Plot (RP)

Dynamic nonlinear systems often exhibit recurrent characteristics, such as oscillations and periodic patterns, that are not always readily discernible from time-domain representations alone. To facilitate the analysis of such behaviors, Eckmann et al. introduced the Recurrence Plot (RP), a two-dimensional visualization technique designed to represent trajectories evolving in a higher-dimensional phase space [3]. An RP representation takes the form of a square matrix and highlights the points at which the system revisits states that are close to previously visited ones in the higher-dimensional phase space.

In this monograph, we adopt the non-binarized formulation of RP proposed in [4], which preserves continuous distance information and avoids the loss of information associated with threshold-based binarization. The construction of an RP representation from a univariate time series involves two main steps. First, an embedding dimension m is selected to reconstruct the higher-dimensional phase space. Using the time-delay embedding technique, an m-dimensional phase space trajectory $\overrightarrow{S} = \{\overrightarrow{s_1}, \overrightarrow{s_2} \ldots, \overrightarrow{s_{L-m+1}}\}$ is formed from the scaled time series $\hat{x} = \{\hat{x}_1, \hat{x}_2, \ldots, \hat{x}_L\}$, where each state vector is defined as $\overrightarrow{s_i} = (\hat{x}_i, \hat{x}_{i+1}, \ldots, \hat{x}_{i+m-1})$.

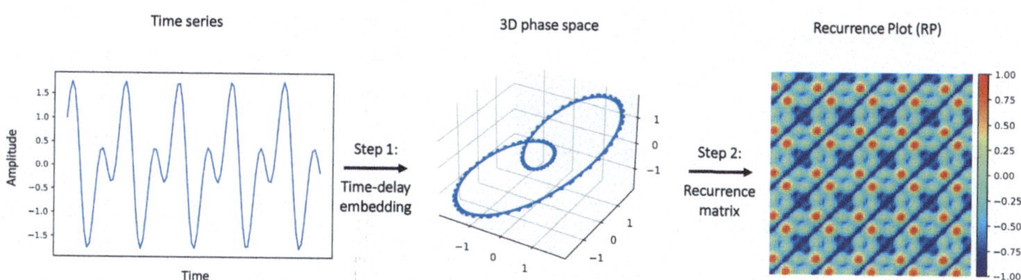

Fig. 3.2 Illustration of the encoding procedure that encodes a univariate time series into a two-dimensional RP representation. Left: a scaled one-dimensional time series $\hat{x} = \{\hat{x}_1, \hat{x}_2, \ldots, \hat{x}_{20}\}$ with $L = 20$ time steps. Middle: reconstruction of the m-dimensional phase space using time-delay embedding, where $m = 3$ and each state is given by $\vec{s_i} = (\hat{x}_i, \hat{x}_{i+1}, \hat{x}_{i+2})$. Right: the resulting RP representation, an 18×18 matrix, computed using Eq. 3.3. Adapted from [2], with permission

In the second step, the RP matrix is computed by evaluating the pairwise distances between all reconstructed states, according to

$$\mathrm{RP}[i, j] = \left\| \vec{s_i} - \vec{s_j} \right\|, \qquad i, j = 1, 2, \ldots, L - m + 1, \tag{3.3}$$

where $\|\cdot\|$ denotes the Euclidean norm. Therefore, each cell (or pixel) of the resulting RP matrix corresponds to the distance between two states in the m-dimensional phase space. Smaller values (or intensities) indicate greater closeness of the states in the m-dimensional phase space, and vice versa. The complete procedure for encoding a univariate time series into an RP representation is shown in Fig. 3.2.

As demonstrated in Fig. 3.3, both the GAF and RP representations exhibit distinctive textures and structural patterns. While GAF images primarily capture information related to the *static* behaviour of a time series, RP representations emphasize *recurrent* behaviour.

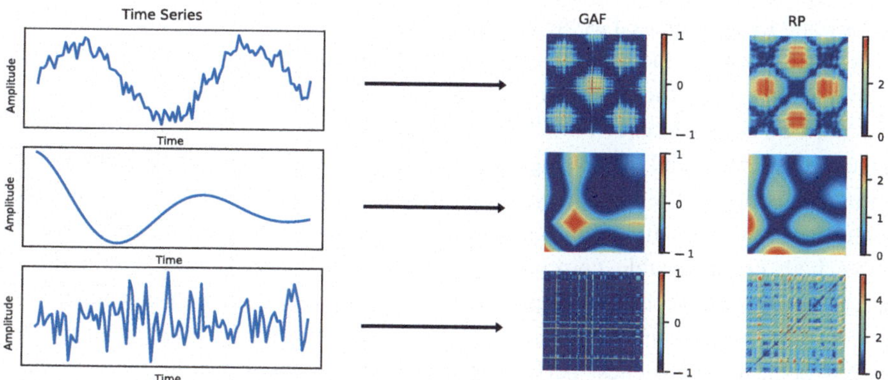

Fig. 3.3 Qualitative comparison of GAF and RP representations for three distinct time series signals, highlighting how each imaging technique reveals different structural characteristics of the underlying temporal dynamics. Adapted from [2], with permission

3.3 Performance Evaluation on Benchmark Datasets

In this section, we evaluate the feature engineering pathway for visual analytics on two benchmark datasets, CSTH and Arc Loss. Specifically, we compare models trained directly on raw time series data with those trained on the corresponding GAF and RP representations. This comparison aims to assess the value of time series imaging as a feature engineering step to improve model prediction performance.

Moreover, we analyze the GAF and RP representations generated from the benchmark datasets. Our objective is to understand the characteristic properties each method captures from the raw time series data. This qualitative analysis helps reveal the temporal and cross-variable correlations present in the visual domain and highlights potential limitations of time series imaging tools.

3.3.1 Quantitative Evaluation: Model Prediction Performance

The experiments conducted in this study involve training, validating, and testing three different models to predict process faults in the two benchmark datasets: CSTH and Arc Loss. The models are: (1) CNN, (2) GAF+CNN, and (3) RP+CNN. For model (1), the raw MTS are reshaped into two-dimensional matrices by stacking the process variables along one dimension and the time steps along the other (refer to Fig. 1.6). For models (2) and (3), the raw data are first converted into GAF and RP representations, respectively, before being input to the CNN. The same CNN architecture and hyperparameters are used for all models to ensure a fair comparison.

A hold-out strategy, consistent with the data split described in Chap. 2, is used for model evaluation. A random search is performed over a predefined hyperparameter space, and the best configuration is selected based on validation performance. The results are reported on the testing set using three metrics: accuracy, F_1 score, and training time (TT). Accuracy and F_1 are reported as the mean and standard deviation across five independent runs, while TT denotes the total training time for 100 epochs on an NVIDIA A100 GPU (51 GB VRAM). The experimental results are summarized in Table 3.1.

For the CSTH benchmark, the CNN model trained on raw data achieves an average accuracy of 97.65% and an average F_1 score of 97.65%. This high fault detection performance is due to the simplicity of the underlying process and the small number of variables ($p = 3$). By employing a kernel size of 3, the CNN receptive field effectively captures both cross-variable correlations and temporal dependencies within the data. However, applying GAF and RP representations as inputs to CNN further improves the fault detection performance. The GAF+CNN model reaches 98.96% accuracy, while RP+CNN achieves the highest performance at 99.62%. These results indicate that the two-dimensional visual encodings capture complex correlations and patterns that are not readily apparent in the raw time series data. These patterns become more apparent in the visual domain. However, these benefits come with higher computational costs; TT

Table 3.1 Performance comparison of CNN, GAF+CNN, and RP+CNN on the CSTH and Arc Loss datasets. Bold values indicate the best performance for each metric

Model	Accuracy (%)	F_1(%)	TT (min)
CSTH			
CNN	97.65 ± 0.89	97.65 ± 0.89	**3.73**
GAF+CNN	98.96 ± 0.37	98.96 ± 0.37	17.14[a]
RP+CNN	**99.62 ± 0.07**	**99.62 ± 0.07**	16.99[a]
Arc Loss			
CNN	**73.81 ± 1.02**	**73.48 ± 0.96**	**31.88**
GAF+CNN	70.66 ± 0.36	69.45 ± 0.86	88.56[a]
RP+CNN	71.66 ± 0.88	70.89 ± 1.18	74.74[a]

[a]TT includes the time required to encode raw data into visual representations

increases from 3.73 minutes for CNN to over 16.99 minutes for both imaging-based models due to the additional computational overhead required to encode raw data into visual representations.

In contrast, for the Arc Loss dataset, the CNN trained on raw data performs best, achieving 73.81% accuracy and 73.48% F_1 score. However, this model has a limitation: local correlations between process variables that fall outside the same receptive field are only partially captured by CNN. This makes the CNN model performance highly dependent on the order in which the variables are arranged in the fused 2D matrix. Specifically, arranging highly correlated process variables closer together would lead to better performance compared to arrangements where these variables are placed farther apart.

The GAF+CNN and RP+CNN models perform slightly worse, with average accuracies of 70.66% and 71.66%, respectively. These results show a performance drop of 3.15% and 2.15% compared to the CNN model trained on raw data. The high dimensionality of the Arc Loss dataset led to the use of tailored preprocessing, such as piecewise aggregate approximation (PAA) [5], to reduce the computational complexity of converting time series into GAF and RP representations. Even though this reduction is necessary for practical reasons, it causes a loss of temporal information in the data. As a result, the GAF and RP representations, built on the smoothed/ reduced time series, are less effective in capturing detailed autocorrelations and temporal patterns present in the original raw time series data.

Moreover, the CNN model trained on raw data is significantly faster, with a TT of 31.88 minutes. In contrast, the GAF+CNN and RP+CNN models require considerably more time, with TT of 88.56 and 74.74 minutes, respectively. This is due to both the preprocessing step using PAA and the computational requirements for generating the visual representations. This higher computational demand highlights a key limitation of using imaging tools in industrial datasets with high dimensionality and long time series lengths.

3.3.2 Qualitative Evaluation: Comparative Analysis of Visual Representations

This subsection presents a qualitative analysis of the GAF and RP representations obtained for the CSTH and Arc Loss benchmark datasets. The objective is to examine the visual patterns embedded within these representations. Since visual interpretability cannot be characterized by a single quantitative metric, a qualitative evaluation approach is adopted to analyze and interpret the information revealed by the GAF and RP encodings.

3.3.2.1 CSTH Benchmark

Figure 3.4 illustrates two normal operating samples ($Y = 0$) from the CSTH dataset and their corresponding GAF and RP representations. The representation are displayed as RGB images, where the red, green, and blue channels correspond to temperature, level, and CW flow measurements, respectively. Examination of the GAF images reveals characteristic "L-shaped" patterns, which are indicative of changes in process set points. Furthermore, homogeneous regions appearing as uniform blocks in the upper-right portions of the GAF images can be observed. These regions correspond to stationary operating conditions during which the system remains in a stable regime for a sustained period. The spatial extent of these homogeneous regions is directly related to the time duration of stationarity. The sequential appearance of an "L-shaped" transition followed by a homogeneous

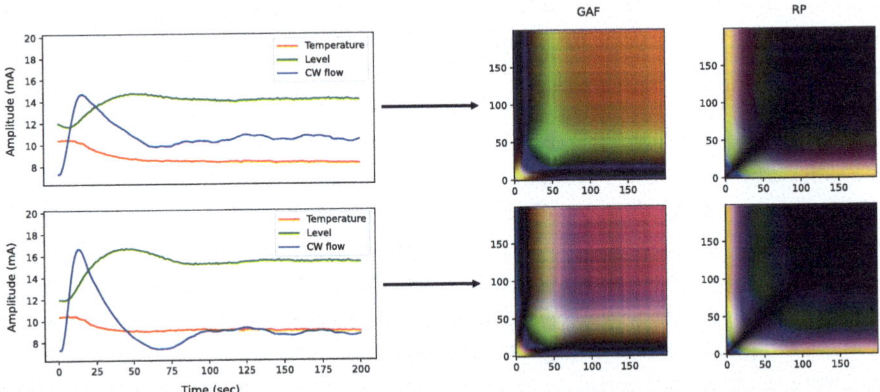

Fig. 3.4 Two normal CSTH samples ($Y = 0$) and their associated GAF and RP representations. Several qualitative observations can be made: (1) a greenish-yellow (green + red) "L-shaped" pattern in the upper GAF representation around the 52-second mark reflects simultaneous shifts in level and temperature set points, with the dominant green hue indicating a larger change in level; (2) a reddish-white (red + green + blue) homogeneous region in the upper GAF representation indicates stationary behavior across temperature, level, and CW flow measurements, with red dominance corresponding to stronger temperature stationarity; (3) the lower GAF representation exhibits a reddish-purple (red + blue) homogeneous region, indicating stable temperature and CW flow measurements; and (4) largely blank RP representations indicate the absence of recurrent dynamics. Adapted from [2], with permission

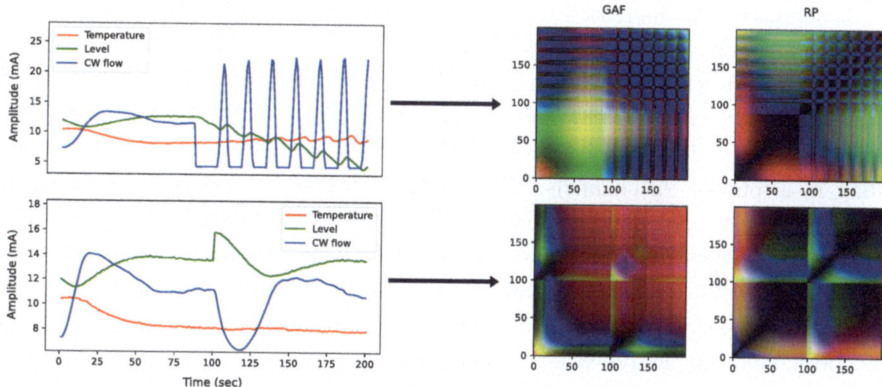

Fig. 3.5 Two faulty CSTH samples ($Y = 1$) and their corresponding GAF and RP representations. Qualitative observations include: (1) distinct periodic patterns in both GAF and RP representations resulting from cyclic system behavior, where the spacing between patterns corresponds to the oscillation period; and (2) bright corner regions in the GAF representations, such as the bright red upper corner in the lower GAF representation, indicating changes in trends associated with increasing or decreasing process variables. Adapted from [2], with permission

region reflects a typical operating pattern in the CSTH process, namely a set-point change followed by steady-state behavior. Such patterns are representative of normal operating conditions in the system. In contrast, the corresponding RP representations are largely devoid of structure, indicating the absence of recurrent or periodic behavior. This observation is consistent with the fact that the CSTH system does not exhibit repeating dynamics under normal conditions.

Figure 3.5 provides a contrasting view by visualizing two faulty CSTH samples ($Y = 1$) alongside their associated GAF and RP representations. Here, both GAF and RP representations display clear periodic structures, reflecting the presence of cyclic behavior in the underlying system. The spacing between repeated patterns corresponds to the temporal period of the underlying oscillations. This behavior is expected, as faults in the simulated CSTH dataset are intentionally designed to induce repeating dynamics. Specifically, when a fault occurs at a given time step, the system, which is otherwise well-controlled, is brought back to a steady state by the controller. This results in two steady states, before and after the fault, forming a recurring pattern separated by the fault event. Thus, the periodic patterns observed in both GAF and RP representations serve as informative indicators of faulty operating regimes.

3.3.2.2 Arc Loss Benchmark

The Arc Loss benchmark introduces additional complexity due to its industrial origin and high dimensionality. In contrast to the CSTH dataset, which comprises only three process variables, the Arc Loss dataset includes 96 variables. Moreover, the time series length in the Arc Loss dataset is 1101 time steps, compared to 200 time steps for the

CSTH benchmark. This increased scale necessitates specialized preprocessing strategies to address both computational and analytical challenges.

Both GAF and RP encodings are inherently designed for univariate time series analysis. Hence, when applied to multivariate systems such as the Arc Loss benchmark, these techniques produce representations with as many channels as process variables, resulting in a total of 96 channels. Visualizing such high-dimensional representations is impractical using conventional display tools, which are typically limited to three-channel (RGB) images. To address this limitation, three key process variables are selected for visualization: total power (TP), furnace feed (FF), and furnace off-gas temperature (FOGET). This selection is informed by domain knowledge, as these variables exhibit strong correlations that are relevant for identifying arc loss events.

Figure 3.6 presents GAF and RP representations, displayed as RGB images, for FOGET, FF, and TP measurements recorded during two normal operating periods ($Y = 0$). In contrast, Fig. 3.7 illustrates the corresponding representations obtained from two faulty operating periods ($Y = 1$).

Despite the informative patterns revealed by GAF and RP representations, several limitations arise when applying them to large-scale industrial datasets. A primary challenge lies in selecting an appropriate subset of variables for visualization, a process that can be time-consuming and labor-intensive in high-dimensional systems, particularly when process knowledge is unavailable. In addition, the use of PAA, while necessary to reduce computational complexity, introduces smoothing effects that may obscure important transient phenomena. For example, isolated abnormal measurements may be averaged out during dimensionality reduction, preventing their manifestation in the resulting GAF or

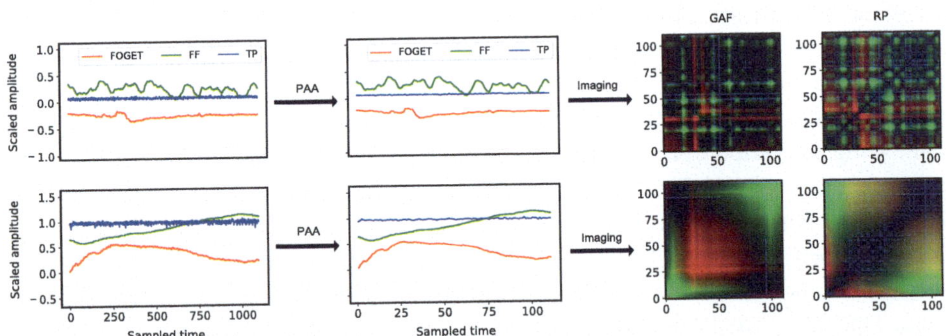

Fig. 3.6 Illustration of two normal Arc Loss samples ($Y = 0$) and their step-by-step transformation into GAF and RP representations. First, three variables (FOGET, TP, and FF) are selected for visualization. The raw time series are then reduced using PAA with a window size of 10, followed by GAF and RP encodings. The resulting representations are displayed as RGB images, with the red, green, and blue channels corresponding to FOGET, FF, and TP, respectively. In the first sample (upper row), homogeneous regions indicate stationary behavior, while red vertical and horizontal lines mark the duration of a disturbance in FOGET measurements. In the second sample, bright green and red corner regions highlight increasing and decreasing trends in FF and FOGET measurements, respectively. Adapted from [2], with permission

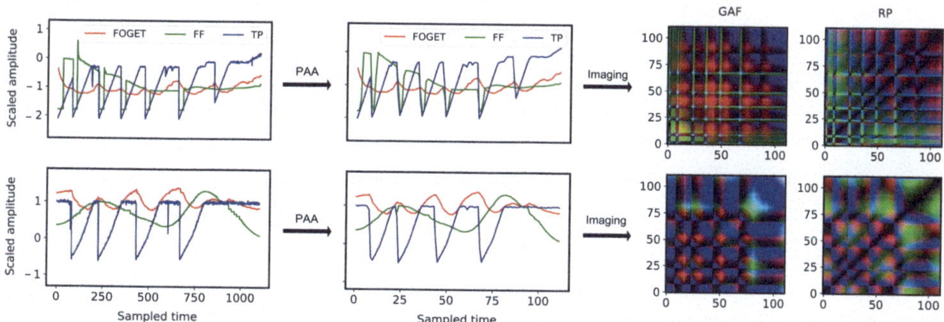

Fig. 3.7 GAF and RP representations displayed as RGB images for FOGET, FF, and TP measurements recorded during two faulty operating periods ($Y = 1$). The appearance of periodic checkerboard-like structures in both representations suggests underlying fluctuations in FOGET, TP, and FF measurements. Pronounced power drops, accompanied by large variations in FF and FOGET, define an arc loss event. Adapted from [2], with permission

RP images. Furthermore, the temporal distances represented in these images correspond to the reduced time scale induced by PAA. As a result, each pixel in the GAF and RP representations reflects the average behavior over two time windows of length 10 in the original signal. This temporal aggregation can mask short-lived dynamics that are critical for understanding industrial process behavior, particularly during the onset of fault conditions.

References

1. Wang Z, Oates T. Imaging time-series to improve classification and imputation. *arXiv preprint arXiv:1506.00327*. 2015. https://arxiv.org/abs/1506.00327
2. Yousef I, Tulsyan A, Shah SL, Gopaluni RB. Visual analytics for process monitoring: Leveraging time-series imaging for enhanced interpretability. *Journal of Process Control*. 2023;132:103127. https://doi.org/10.1016/j.jprocont.2023.103127
3. Eckmann JP, Kamphorst SO, Ruelle D. Recurrence plots of dynamical systems. *Europhysics Letters (EPL)*. 1987;4(9):973–977. https://doi.org/10.1209/0295-5075/4/9/004
4. Souza VMA, Silva DF, Batista GEAPA. Extracting texture features for time series classification. In: *Proceedings of the 22nd International Conference on Pattern Recognition*. IEEE; 2014. p. 1425–1430. https://doi.org/10.1109/ICPR.2014.254
5. Keogh E, Chakrabarti K, Pazzani M, Mehrotra S. Dimensionality reduction for fast similarity search in large time series databases. *Knowledge and Information Systems*. 2001;3(3):263–286. https://doi.org/10.1007/PL00011669

Visual Analytics Pathway II: Architecture Engineering

4

Contents

4.1 Background and Related Work

In this section, we begin by introducing the fundamental concepts and definitions relevant to our study. Following that, we provide an overview of time series classifiers and discuss the basics of convolution operations, which serve as the primary component in CNN.

4.1.1 Overview of Time Series Classification Methods

In the context of FDD, the objective of TSC is to correctly assign input process data, in the form of time series, to their corresponding operating conditions. Broadly speaking, existing TSC methods can be categorized into three main groups according to their underlying algorithmic principles: distance-based, feature-based, and deep learning methods.

Distance-based methods assess the similarity or dissimilarity between pairs of time series using predefined distance measures [1]. Considerable research has focused on the development of *elastic* distance measures that can accommodate small temporal misalignments between signals [2]. These measures improve robustness when comparing

I. Yousef et al., *Visual Analytics for Process Monitoring*, Synthesis Lectures on Emerging Engineering Technologies, https://doi.org/10.1007/978-3-032-22126-1_4

time series that exhibit phase shifts or local temporal distortions by allowing flexible alignment of time steps. Among elastic distance measures, dynamic time warping (DTW) is one of the most widely adopted techniques [3]. In fact, DTW, followed by 1-nearest neighbor classifier (DTW+1NN) is commonly regarded as a benchmark approach, often referred to as the *gold standard* in the literature [4].

Feature-based methods constitute the second class of TSC approaches and are based on the extraction of informative features from raw time series data [5]. These methods can be broadly divided into interval-based and dictionary-based techniques. Interval-based approaches operate by selecting subsequences from the raw time series and extracting discriminative statistical features over these intervals. A representative example is the time series forest (TSF) algorithm [6]. TSF constructs an ensemble of decision trees using randomly sampled intervals and their associated summary statistics, such as the mean, slope, and standard deviation.

Next, dictionary-based family transforms time series into symbolic representations and characterize them using the frequency of symbolic patterns [7]. This is typically achieved by discretizing the signal, extracting words through a sliding window mechanism, and constructing a histogram over a predefined dictionary. The bag of symbolic Fourier approximation symbols (BOSS) algorithm is a prominent method within this family [8]. BOSS represents time series as bags of words derived from symbolic Fourier approximation (SFA) coefficients. The SFA technique converts time series into compact symbolic forms by approximating their Fourier transforms [9]. A dictionary of SFA words is learned from the training data, and each time series is subsequently encoded as a histogram that captures the occurrence frequencies of these words. Classification is then performed using similarity or distance measures between histogram representations.

Thirdly, deep learning approaches form the third category of TSC methods and are based on neural networks that learn hierarchical representations directly from time series data [10]. A neural network is typically described as *deep* when it contains more than one hidden layer between the input and output layers. More formally, a deep neural network consists of K parametric layers, where each layer learns progressively more abstract representations of the input space [11]. One of the simplest deep learning architectures is the multi-layer perceptron (MLP), also referred to as a fully connected network (FCN) or artificial neural network (ANN) [12]. In an MLP, each neuron in layer k_i is fully connected to neurons in layers k_{i-1} and k_{i+1}, for $i \in [2, K - 1]$. These connections are modelled by learnable weights within the neural network.

A key limitation of MLP-based models for TSC is their lack of temporal or spatial invariance. Specifically, each time step is associated with a distinct set of weights, which limits their ability to capture sequential structure. To better model temporal dependencies, recurrent neural networks (RNNs) were introduced [13, 14]. RNNs maintain a hidden state that evolves over time, which enables them to incorporate information from previous time steps. However, standard RNN architectures are known to suffer from vanishing and exploding gradient problems, which restrict their capacity to learn long-term dependencies [15]. Long short-term memory (LSTM) is a variant of RNN that overcomes the limitations

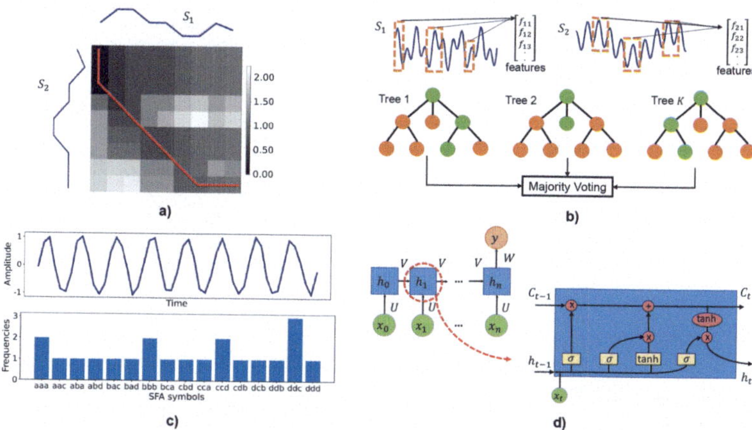

Fig. 4.1 Conceptual overview of representative TSC methods considered in this chapter: (**a**) DTW+1NN, which evaluates similarity between time series using dynamic time warping; (**b**) TSF, which applies a random forest classifier to features extracted from randomly selected time intervals; (**c**) BOSS, which represents time series as histograms of symbolic Fourier approximation words; and (**d**) LSTM, which captures temporal dependencies through a gated memory mechanism

of vanilla RNN [16]. Its architecture includes a cell state, input gate, forget gate, and output gate. This architecture allows LSTMs to selectively retain or discard information over extended sequences while mitigating gradient-related challenges [17, 18]. As a result, LSTM models have been successfully applied to sequential learning tasks, including TSC and FDD [19, 20].

In this chapter, the proposed visual analytics framework is benchmarked against representative methods from each of the three TSC categories: (1) distance-based methods, represented by DTW+1NN; (2) feature-based methods, represented by TSF and BOSS; and (3) deep learning methods, represented by LSTM. A high-level conceptual comparison of these classifiers is provided in Fig. 4.1.

4.1.2 Convolutional Feature Extraction

Convolution operations are the core mechanism used by CNN to extract features from input data [21]. A convolution operation involves an input signal and a kernel (an operator function). Conceptually, convolution can be viewed as a mathematical transformation that highlights and extracts relevant features from the input data [22]. Convolution can be applied to signals of different dimensions, such as 1D, 2D, or 3D, with the kernel dimensions matching those of the input. In practice, 1D convolution is commonly used for processing audio signals [23, 24], while 2D convolution is widely adopted for image analysis tasks [25, 26]. Similarly, 3D convolution is often applied to video data [27]. In this work, the focus is limited to 1D and 2D convolutions, as no video data are considered.

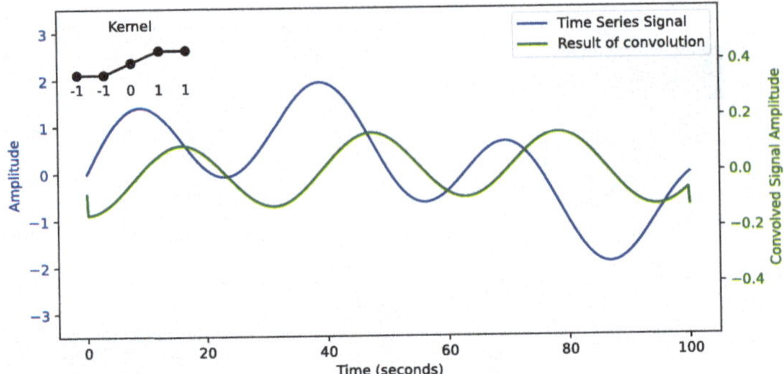

Fig. 4.2 An illustration of a 1D convolution operation. The gradient kernel, [−1, −1, 0, 1, 1], is applied to a 1D signal to extract changes in amplitude or slope. The resulting convolved signal highlights regions where the input exhibits positive and negative gradients, providing insight into the overall trend and direction of changes in the original time series

Input matrix

1	1	-1	1	-1	1
-1	0	2	0	2	4
0	1	2	0	3	1
2	2	-1	1	0	-2
1	2	-1	0	1	-1
1	-1	0	0	1	0

Kernel

1	1	-1
-1	1	2
0	1	2

= 13

= 3

Fig. 4.3 An illustration of a 2D convolution operation. When the input patch aligns with the kernel, the convolution produces a large activation

In a convolution operation, a kernel slides over the input data and performs a dot product to generate features known as feature maps [28]. The kernel, also known as a filter or convolutional filter, is a set of weights that defines how the input is transformed. The design of the kernel determines which types of features are extracted. These features represent hidden patterns in the data, such as trends or variations that may not be apparent. The kernel is typically smaller than the input, which reduces computational cost. In 1D convolution, the kernel is a weight vector, and the feature map is obtained by adding a bias term to the sliding dot product between the 1D input and the kernel weights [29]. Figure 4.2 shows an example of a 1D convolution operation using a gradient kernel applied to a time series signal.

Next, 2D convolutions are applied to two-dimensional (2D) data (e.g., images), where the kernel is a matrix of weights. The resulting feature maps indicate the degree of similarity between the input and the kernel pattern. Figure 4.3 illustrates a 2D convolution operation. Convolution operations share similarities with feature-based approaches, as both depend on identifying specific patterns or motifs within the input data. In CNN, multiple kernels with different weights are used to capture a variety of patterns and variations within the input data. This combination of kernels allows for the detection of complex and discriminative features. The success of CNN in time series and image classification demonstrates the effectiveness of convolutional kernels as the foundation for extracting meaningful representations from raw data.

4.2 Architecture of the Proposed Framework

This chapter is inspired by our previous studies on visual analytics for process monitoring, where process data, in the form of MTS, are transformed into discriminative visual representations for FDD [30, 31]. In this section, we present an end-to-end visual analytics framework for FDD. The architecture, illustrated in Fig. 4.4, is composed of four main stages that operate sequentially to extract, visualize, and classify temporal and spatial features in a supervised setting. The main objective of the proposed network is to maximize the visual distinction between representations of samples across different classes.

First, the input $X \in \mathbb{R}^{L\times p}$ is passed through a 1D convolution layer to extract low-level temporal dependencies among process variables. This initial layer transforms the input data X into an intermediate representation for subsequent layers. Next, we use a series of *bottleneck* 1D residual blocks, consisting of three convolutional layers: $1\times1 \rightarrow 1\times3 \rightarrow 1\times1$, with layer normalization layers and rectified linear unit (ReLU) activations (refer to Fig. 4.5). This design stabilizes optimization, improves gradient flow, and enables deeper temporal feature learning with modest parameter cost. The number of blocks and channels per block are hyperparameters; channels typically increase across blocks to encode progressively higher-level patterns. We use three residual blocks and omit pooling layers in the network to preserve valuable spatial information. We also apply "same" padding and a stride size set to 1. Stage 1 of the network seeks to learn discriminative visual representations using 1D convolution operations.

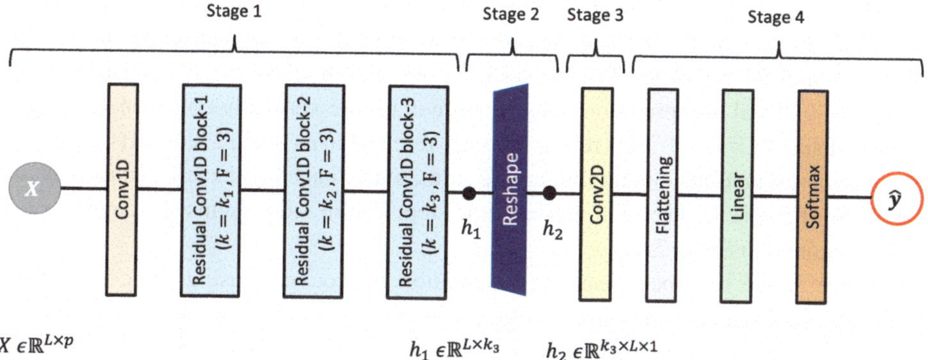

Fig. 4.4 The proposed network architecture. The residual block configuration is presented in Fig. 4.5. Abbreviations: k—number of kernels, F—kernel size, L—the size of the input signal (i.e. the number of time steps), p—number of variables, $\hat{y}$—predicted class

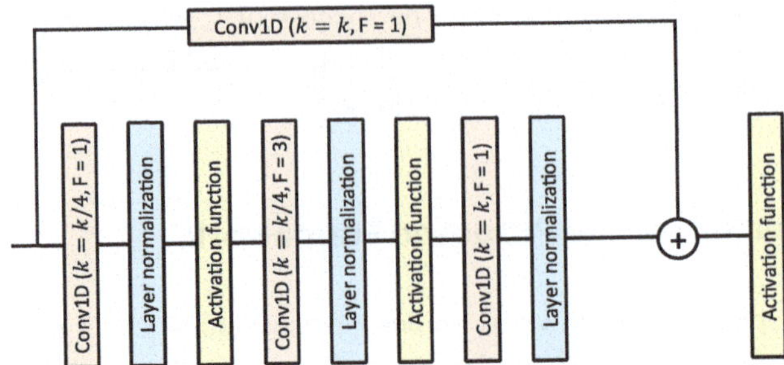

Fig. 4.5 The residual block configuration. Abbreviations: k—number of kernels, F—kernel size

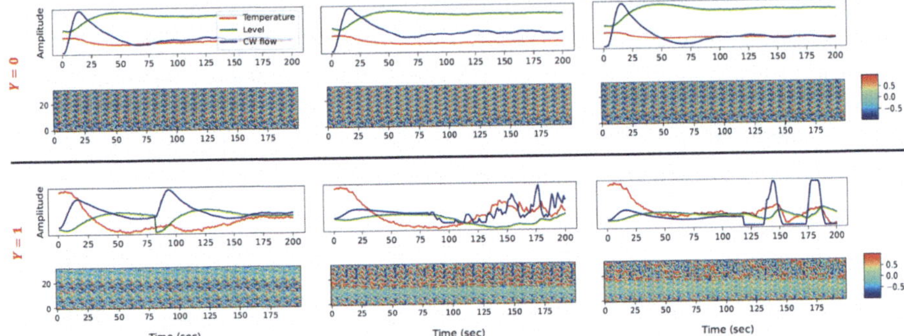

Fig. 4.6 Visual comparison of the representations obtained using the proposed visual analytics framework for three normal (top) and three faulty (bottom) samples from the CSTH benchmark. Homogeneous textures are associated with normal operating conditions, whereas irregular textures indicate faulty behaviour. Normal and faulty signals respond differently to the learned one-dimensional convolutional kernels, resulting in visually distinct representations

The output of stage 1 is a combination of k_3 feature maps, corresponding to the number of 1D kernels in the last residual block of size L (assuming that "same" convolution is employed). Each feature map captures the activation level of its corresponding kernel, which indicates the response of that specific kernel pattern across the features of the input X. To enable visual interpretation, these feature maps are vertically concatenated to form an array of size $k_3 \times L \times 1$. This reshaping step, referred to as Stage 2, converts the learned temporal representations into an image-like form, where each pixel encodes the activation magnitude of a specific kernel at a given time step. Figure 4.6 presents six samples of the images obtained in Stage 2.

In Stage 3, the reshaped visual representation is processed using 2D convolutional layers to extract spatial correlations among features. Each 2D convolution block consists of a convolution layer, followed by a layer normalization layer, and a nonlinear activation function. "Valid" padding is used to progressively reduce the spatial dimensions of the feature maps, therefore improving computational efficiency. The 2D convolution operations learn to identify spatial patterns that characterize distinct operating conditions, which allow the network to uncover relationships not apparent in the raw temporal domain.

Finally, Stage 4 performs classification using an MLP. The feature maps obtained from the last 2D convolution block are flattened into a one-dimensional vector, which is then passed through fully connected layers. Flattening is the process of converting the multidimensional feature maps into a one-dimensional vector. The MLP network consists of multiple dense layers that map the extracted visual features into a scalar that represents a class label. The output layer employs a softmax activation function, which produces a probability distribution over the classes. The model parameters are optimized by minimizing the cross-entropy loss function, defined as:

$$\mathcal{L} = -\frac{1}{N} \sum_{i=1}^{N} \left[y_i \log(\hat{y}_i) + (1 - y_i) \log(1 - \hat{y}_i) \right], \tag{4.1}$$

where y_i and $\hat{y}_i$ denote the true and predicted class labels for the i-th sample, respectively, and N is the total number of samples. Dropout regularization is applied to prevent overfitting by randomly deactivating neurons during training [32].

4.3 Performance Evaluation on Benchmark Datasets

In this section, we evaluate our proposed visual analytics approach using two benchmark datasets: CSTH and Arc Loss. We begin with the simulated CSTH benchmark, which provides a controlled setting to demonstrate the applicability and benefits of our approach. By analyzing the simulated data, we highlight the interpretability and insights gained through our visual analytics framework. Subsequently, we apply the proposed approach to the industrial Arc Loss benchmark, a large-scale dataset, to test the scalability and robustness of the proposed approach. The performance of the proposed approach is compared with DTW+1NN, TSF, BOSS, LSTM, GAF followed by CNN, and RP followed by CNN models.

4.3.1 Quantitative Evaluation: Model Prediction Performance

For model evaluation, we employ a hold-out strategy. Hyperparameter tuning is performed via a random search over a manually predefined search space. For each sampled configuration, the model is trained on the training subset and evaluated on the validation subset. The configuration achieving the best validation performance is then retrained using the entire training set. To avoid overfitting, an early stopping criterion is applied: if the validation loss does not improve for ten consecutive epochs, training is stopped. The final models are evaluated on the unseen testing sets, and the results are reported as the mean and standard

Table 4.1 Comparative performance on the CSTH and Arc Loss benchmark datasets. Bold values indicate the best performance for each metric

Model	ACC (%)	PPV (%)	TPR (%)	1-FPR (%)	F_1 (%)
CSTH					
DTW+1NN	81.17 ± 0.00	96.38 ± 0.00	64.89 ± 0.00	97.55 ± 0.00	77.56 ± 0.00
TSF	99.14 ± 0.09	99.20 ± 0.13	99.09 ± 0.11	99.20 ± 0.13	99.15 ± 0.09
BOSS	93.89 ± 0.00	94.01 ± 0.00	93.80 ± 0.00	93.98 ± 0.00	93.90 ± 0.00
LSTM	91.23 ± 7.05	90.53 ± 8.76	92.84 ± 4.54	89.60 ± 10.16	91.19 ± 7.10
GAF+CNN	99.15 ± 0.09	99.80 ± 0.13	98.51 ± 0.09	99.80 ± 0.13	99.15 ± 0.09
RP+CNN	**99.71 ± 0.07**	**99.96 ± 0.09**	**99.47 ± 0.18**	**99.96 ± 0.09**	**99.71 ± 0.07**
Proposed	99.22 ± 0.13	99.38 ± 0.44	99.07 ± 0.29	99.37 ± 0.45	99.22 ± 0.13
Arc Loss					
DTW+1NN	64.96 ± 0.00	65.52 ± 0.00	64.31 ± 0.00	65.63 ± 0.00	64.91 ± 0.00
TSF	73.52 ± 0.25	66.40 ± 0.18	96.38 ± 0.30	50.16 ± 0.34	78.63 ± 0.20
BOSS	65.58 ± 0.00	59.67 ± 0.00	**98.47 ± 0.00**	31.97 ± 0.00	74.31 ± 0.00
LSTM	75.50 ± 1.28	73.47 ± 1.30	80.62 ± 2.95	70.38 ± 2.48	76.84 ± 1.45
GAF+CNN	69.06 ± 1.01	64.78 ± 1.57	85.57 ± 8.13	52.21 ± 0.08	73.49 ± 2.24
RP+CNN	71.66 ± 0.88	67.11 ± 1.45	86.32 ± 4.05	56.72 ± 4.78	75.43 ± 0.98
Proposed	**78.60 ± 0.33**	**74.40 ± 1.00**	88.00 ± 3.08	**68.90 ± 2.53**	**80.60 ± 0.72**

deviation across five independent runs with different random seeds using the same data split.

Table 4.1 summarizes the comparative performance of all models across the CSTH and Arc Loss benchmarks using accuracy (ACC), precision (PPV), recall (TPR), 1-false positive rate (FPR), and F_1-score.

4.3.1.1 CSTH Benchmark

On the CSTH benchmark, the proposed visual analytics framework demonstrates competitive performance, ranking second after RP+CNN, which achieves the highest scores across all metrics. The superior performance of RP+CNN can be attributed to the recurrent characteristics of the CSTH dataset. As discussed in Sect. 2.2, the CSTH dataset is simulated so that faults exhibit recurring patterns. The periodic patterns are effectively captured by RP encodings, which makes RP+CNN particularly effective in this scenario. Similarly, GAF encodings are designed to extract static features from the underlying time series. Both RP and GAF are feature extraction methods that yield exceptional results when the underlying system exhibits characteristics that match their feature extraction strengths.

On the other hand, the proposed framework does not rely on predefined feature extraction techniques; instead, it learns data-specific features optimized to maximize the separation between classes. This property is particularly valuable in applications where domain-specific process knowledge is limited or unavailable. Furthermore, DTW+1NN ranks as the least-performing model. This is because faulty samples with minor fault magnitudes are very similar to normal samples, which makes it difficult for DTW+1NN to differentiate between the two classes.

4.3.1.2 Arc Loss Benchmark

On the Arc Loss benchmark, the proposed approach achieves the best performance in four of the five evaluation metrics, while BOSS achieves the highest recall. The relatively lower accuracy of GAF+CNN and RP+CNN is attributed to the use of PAA during preprocessing. PAA reduces temporal fidelity, which is essential for CNN (i.e., downstream classifier) to capture meaningful patterns. As a result, their classification performance, while still competitive, is somewhat compromised. In contrast, our proposed approach scales efficiently, with relatively short training times. This efficiency is attributed to the use of 1D convolutions, which are computationally efficient and well-suited for processing large-scale industrial time series data (e.g., the Arc Loss benchmark dataset). This scalability makes the proposed framework particularly suitable for real-world process monitoring scenarios where computational resources are constrained.

4.3.2 Qualitative Evaluation: Comparative Analysis of Visual Representations

This subsection demonstrates how the proposed visual analytics framework combines predictive performance with visual interpretability. We show how the proposed visual analytics approach enables process operators to visually examine and validate the relationship between model predictions and the underlying process conditions. Since visual interpretability cannot be measured using a single quantitative metric, a qualitative evaluation methodology is adopted to analyze the representations generated by the proposed approach.

4.3.2.1 CSTH Benchmark

Figure 4.6 presents a visual comparison of the representations produced by the proposed approach for three normal and three faulty CSTH samples. In contrast to RP and GAF encodings, which yield multi-channel visual representations, the proposed framework generates a single-channel matrix that can be visualized using an appropriate colour mapping. This mapping assigns colours to matrix values, thereby facilitating visual interpretation. Clear visual differences can be observed between the representations of normal and faulty samples. Specifically, normal samples are characterized by relatively uniform textures, whereas faulty samples exhibit irregular and uneven patterns. These distinctions arise from the application of one-dimensional convolution operations, through which discriminative patterns are extracted directly from the raw time series. As a result, normal and faulty signals produce different responses when convolved with the learned kernels, leading to the observed contrast in the resulting visual representations.

4.3.2.2 Arc Loss Benchmark

Figure 4.7 compares two normal Arc Loss samples ($Y = 0$) and their corresponding visual representations generated by the proposed approach with two faulty samples ($Y = 1$) and their associated representations. Each visual representation has a size of $256 \times 1101 \times 1$, where the width of the grayscale image corresponds to the temporal length of the input signal (i.e., 1101 time steps), and the height reflects the number of learned representation dimensions, equal to the number of one-dimensional convolutional kernels in the final residual block. Although the overall texture differs between normal and faulty samples, fine-grained local patterns are less visually prominent than those observed in RP and GAF representations due to the higher spatial resolution of the visual representations (see Figs. 3.6 and 3.7). As shown in Fig. 4.7, normal operating conditions result in more regular texture patterns, whereas irregular and less smooth patterns characterize faulty conditions.

Compared with traditional imaging techniques such as RP and GAF, the proposed approach offers several practical advantages. First, it operates directly on multivariate time series data and produces single-channel visual representations, which simplifies interpretation and analysis. Second, the framework can process multivariate signals of arbitrary length without requiring dimensionality reduction techniques such as PAA. This capability is enabled by the computational efficiency of one-dimensional convolution operations, which scale linearly with input length. Hence, the proposed approach can effectively operate on large-scale industrial datasets without preprocessing steps that may otherwise introduce information loss or reduce temporal resolution.

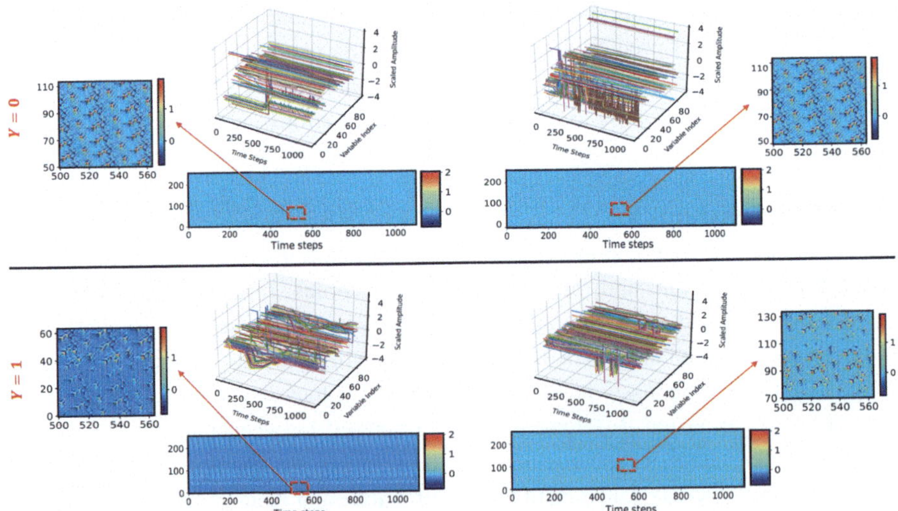

Fig. 4.7 Visual comparison of two normal ($Y = 0$) and two faulty ($Y = 1$) Arc Loss samples obtained using the proposed visual analytics framework. Normal samples exhibit homogeneous texture patterns, whereas faulty samples display irregular and non-smooth textures. Zoomed-in regions highlight local patterns

4.3.3 Sensitivity of Visual Representations to Fault Magnitude

While the previous subsections have demonstrated that the proposed framework can distinguish between normal and faulty operating conditions through visually distinct representations, an important question remains: can the approach reliably capture variations in fault severity, particularly for small-magnitude faults? This subsection investigates the sensitivity of the proposed method to small changes in fault magnitude. The aim is to demonstrate that the framework can detect not only significant deviations from normal operation but also subtle faults, and that these differences are reflected in the resulting visual representations. We compare the visual representation obtained using the proposed approach with those obtained using GAF and RP encodings.

To evaluate sensitivity to fault magnitude, faults of varying sizes are artificially introduced into a normal CSTH sample, and the resulting visual representations are qualitatively examined. Figure 4.8 presents the GAF and RP representations for a normal sample, a faulty sample with a small fault magnitude (5% deviation from the steady-state value), and a faulty sample with a large fault magnitude (30% deviation from the steady-state value), alongside the corresponding time-domain signals. Figure 4.9 illustrates the visual representations of the same samples obtained using the proposed approach.

Across all methods, noticeable visual differences are observed between the normal sample and the faulty sample with a large fault magnitude. However, for faults of small magnitude, the proposed approach produces more distinct visual patterns than GAF and

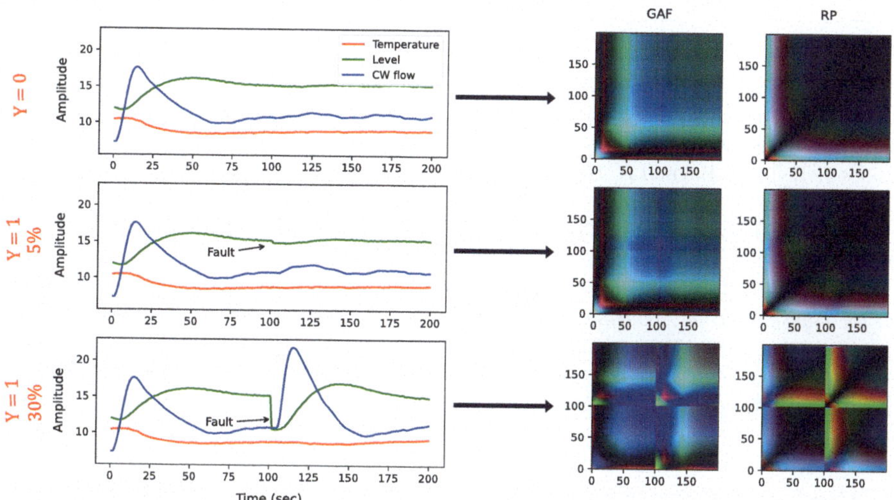

Fig. 4.8 Response of GAF and RP representations to different fault magnitudes. Left: original time-domain signals. Middle: corresponding GAF encodings. Right: corresponding RP encodings. The top row shows a normal CSTH sample, the middle row depicts a faulty sample with a small fault magnitude (5% deviation from steady-state), and the bottom row illustrates a faulty sample with a large fault magnitude (30% deviation from steady-state)

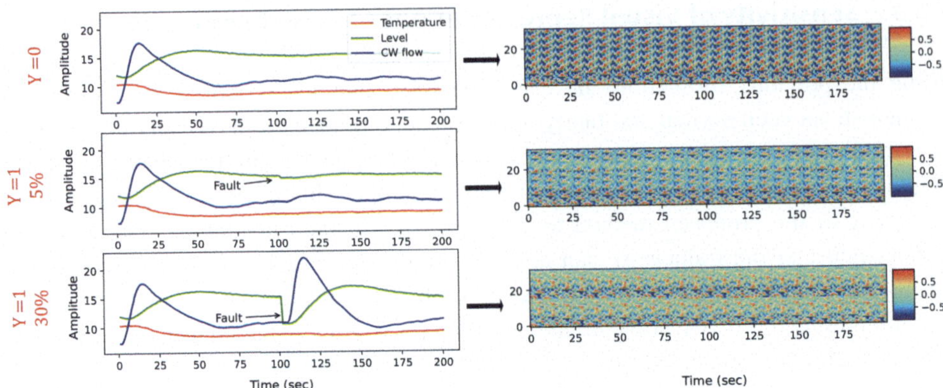

Fig. 4.9 Visual representations generated by the proposed visual analytics framework for a normal CSTH sample (top), a faulty sample with a small fault magnitude of 5% (middle), and a faulty sample with a large fault magnitude of 30% (bottom). Increasing texture irregularity reflects increasing fault severity

RP. Specifically, the visual representations generated by the proposed approach exhibit increasing textural irregularity as fault magnitude increases. This monotonic relationship between texture variation and fault severity indicates that the proposed framework is sensitive to changes in fault magnitude and can capture both minor and severe deviations from normal operating conditions.

References

1. Abanda A, Mori U, Lozano J. A review on distance based time series classification. *Data Mining and Knowledge Discovery*. 2019;33:378–412. https://doi.org/10.1007/s10618-018-0596-4
2. Lines J, Taylor S, Bagnall A. HIVE-COTE: The hierarchical vote collective of transformation-based ensembles for time series classification. In: *Proceedings of the 16th IEEE International Conference on Data Mining (ICDM)*. 2016. p. 1041–1046. https://doi.org/10.1109/ICDM.2016.0133
3. Sakoe H, Chiba S. Dynamic programming algorithm optimization for spoken word recognition. *IEEE Transactions on Acoustics, Speech, and Signal Processing*. 1978;26(1):43–49. https://doi.org/10.1109/TASSP.1978.1163055
4. Lines J, Bagnall A. Time series classification with ensembles of elastic distance measures. *Data Mining and Knowledge Discovery*. 2015;29(3):565–592. https://doi.org/10.1007/s10618-014-0361-2
5. Lei Y, Wu Z. Time series classification based on statistical features. *EURASIP Journal on Wireless Communications and Networking*. 2020;2020:46. https://doi.org/10.1186/s13638-020-1661-4
6. Deng H, Runger G, Tuv E, Vladimir M. A time series forest for classification and feature extraction. *Information Sciences*. 2013;239:142–153. https://doi.org/10.1016/j.ins.2013.02.030
7. Large J, Bagnall A, Malinowski S, Tavenard R. On time series classification with dictionary-based classifiers. *Intelligent Data Analysis*. 2019;23:1073–1089. https://doi.org/10.3233/IDA-184333

8. Schäfer P. The BOSS is concerned with time series classification in the presence of noise. *Data Mining and Knowledge Discovery*. 2015;29:1505–1530. https://doi.org/10.1007/s10618-014-0377-7
9. Schäfer P, Högqvist M. SFA: A symbolic Fourier approximation and index for similarity search in high dimensional datasets. In: *Proceedings of the 15th International Conference on Extending Database Technology*. ACM; 2012. p. 516–527. https://doi.org/10.1145/2247596.2247656
10. Schmidhuber J. Deep learning in neural networks: An overview. *Neural Networks*. 2015;61:85–117. https://doi.org/10.1016/j.neunet.2014.09.003
11. Papernot N, McDaniel P. Deep k-nearest neighbors: Towards confident, interpretable and robust deep learning. *arXiv preprint arXiv:1803.04765*. 2018. https://arxiv.org/abs/1803.04765
12. Wang Z, Yan W, Oates T. Time series classification from scratch with deep neural networks: A strong baseline. In: *Proceedings of the International Joint Conference on Neural Networks (IJCNN)*. 2017. p. 1578–1585. https://doi.org/10.1109/IJCNN.2017.7966039
13. Sherstinsky A. Fundamentals of recurrent neural network (RNN) and long short-term memory (LSTM) network. *Physica D: Nonlinear Phenomena*. 2020;404:132306. https://doi.org/10.1016/j.physd.2019.132306
14. Schmidt RM. Recurrent neural networks (RNNs): A gentle introduction and overview. *arXiv preprint arXiv:1912.05911*. 2019. https://arxiv.org/abs/1912.05911
15. Pascanu R, Mikolov T, Bengio Y. On the difficulty of training recurrent neural networks. *arXiv preprint arXiv:1211.5063*. 2013. https://arxiv.org/abs/1211.5063
16. Hochreiter S, Schmidhuber J. Long short-term memory. *Neural Computation*. 1997;9(8):1735–1780. https://doi.org/10.1162/neco.1997.9.8.1735
17. Van Houdt G, Mosquera C, Nápoles G. A review on the long short-term memory model. *Artificial Intelligence Review*. 2020;53(8):5929–5955. https://doi.org/10.1007/s10462-020-09838-1
18. Yu Y, Si X, Hu C, Zhang J. A review of recurrent neural networks: LSTM cells and network architectures. *Neural Computation*. 2019;31(7):1235–1270. https://doi.org/10.1162/neco_a_01199
19. Yang R, Huang M, Lu Q, Zhong M. Rotating machinery fault diagnosis using long-short-term memory recurrent neural network. *IFAC-PapersOnLine*. 2018;51(24):228–232. https://doi.org/10.1016/j.ifacol.2018.09.582
20. Belagoune S, Bali N, Bakdi A, Baadji B, Atif K. Deep learning through LSTM classification and regression for transmission line fault detection, diagnosis and location in large-scale multi-machine power systems. *Measurement*. 2021;177:109330. https://doi.org/10.1016/j.measurement.2021.109330
21. Yamashita R, Nishio M, Do R, Togashi K. Convolutional neural networks: An overview and application in radiology. *Insights into Imaging*. 2018;9:611–629. https://doi.org/10.1007/s13244-018-0639-9
22. Ajit A, Acharya K, Samanta A. A review of convolutional neural networks. In: *Proceedings of the International Conference on Emerging Trends in Information Technology and Engineering (ic-ETITE)*. 2020. p. 1–5. https://doi.org/10.1109/ic-ETITE47903.2020.049
23. Li F, Liu M, Zhao Y, Kong L, Dong L, Liu X, Hui M. Feature extraction and classification of heart sound using 1D convolutional neural networks. *EURASIP Journal on Applied Signal Processing*. 2019;2019:59. https://doi.org/10.1186/s13634-019-0651-3
24. Abdoli S, Cardinal P, Lameiras Koerich A. End-to-end environmental sound classification using a 1D convolutional neural network. *Expert Systems with Applications*. 2019;136:252–263. https://doi.org/10.1016/j.eswa.2019.06.040
25. Szegedy C, Liu W, Jia Y, Sermanet P, Reed S, Anguelov D, Erhan D, Vanhoucke V, Rabinovich A. Going deeper with convolutions. In: *Proceedings of the IEEE Conference on Computer Vision and Pattern Recognition (CVPR)*. 2015. p. 1–9. https://doi.org/10.1109/CVPR.2015.7298594

26. He K, Zhang X, Ren S, Sun J. Deep residual learning for image recognition. In: *Proceedings of the IEEE Conference on Computer Vision and Pattern Recognition (CVPR)*. 2016. p. 770–778. https://doi.org/10.1109/CVPR.2016.90
27. Karpathy A, Toderici G, Shetty S, Leung T, Sukthankar R, Fei-Fei L. Large-scale video classification with convolutional neural networks. In: *Proceedings of the IEEE Conference on Computer Vision and Pattern Recognition (CVPR)*. 2014. p. 1725–1732. https://doi.org/10.1109/CVPR.2014.223
28. O'Shea K, Nash R. An introduction to convolutional neural networks. *arXiv preprint arXiv:1511.08458*. 2015. https://arxiv.org/abs/1511.08458
29. Kiranyaz S, Avci O, Abdeljaber O, Ince T, Gabbouj M, Inman DJ. 1D convolutional neural networks and applications: A survey. *Mechanical Systems and Signal Processing*. 2021;151:107398. https://doi.org/10.1016/j.ymssp.2020.107398
30. Yousef I, Shah SL, Gopaluni RB. Visual analytics: A new paradigm for process monitoring. *IFAC-PapersOnLine*. 2022;55(7):376–383. https://doi.org/10.1016/j.ifacol.2022.07.473
31. Yousef I, Tulsyan A, Shah SL, Gopaluni RB. Visual analytics for process monitoring: Leveraging time-series imaging for enhanced interpretability. *Journal of Process Control*. 2023;132:103127. https://doi.org/10.1016/j.jprocont.2023.103127
32. Srivastava N, Hinton G, Krizhevsky A, Sutskever I, Salakhutdinov R. Dropout: A simple way to prevent neural networks from overfitting. *Journal of Machine Learning Research*. 2014;15(56):1929–1958.

Visual Analytics Pathway III: Data Engineering

5

Contents

5.1 Background and Related Work

This section introduces the self-supervised learning (SSL) paradigm, with a focus on contrastive learning approaches for time series. It outlines state-of-the-art methods and their underlying principles, emphasizing how they address temporal dependencies. The discussion then highlights the limitations of existing approaches that motivate the development of the proposed CDPCC framework and concludes with an intuitive rationale for its design.

Self-supervised Learning (SSL)

SSL is a recently popular learning paradigm in which models are trained on predictive tasks derived directly from raw data, without requiring manually annotated labels. Although SSL has only recently gained widespread attention, its conceptual foundations can be traced to earlier work that was initially categorized under *unsupervised learning* [1]. Several influential research efforts that are now regarded as foundational for SSL, including DeepCluster [2], Instance Discrimination [3], and context prediction [4], were originally introduced as unsupervised approaches. The recent rebranding of these methods

I. Yousef et al., *Visual Analytics for Process Monitoring*, Synthesis Lectures on Emerging Engineering Technologies, https://doi.org/10.1007/978-3-032-22126-1_5

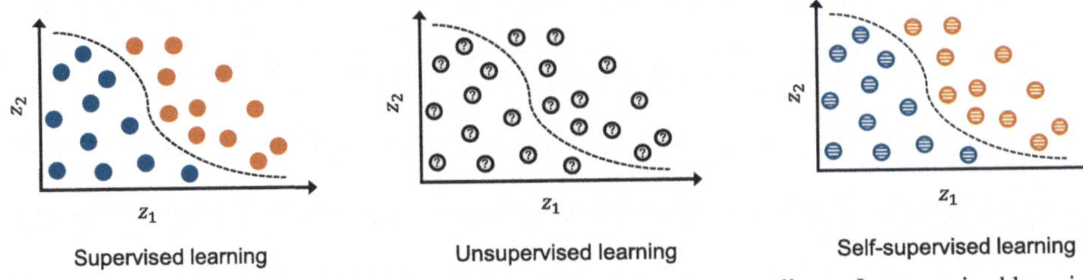

Fig. 5.1 Conceptual comparison of three representation learning paradigms. In supervised learning (left), models use class labels (blue and orange circles) to separate samples from different classes. In unsupervised learning (middle), models identify latent structures within unlabeled data points (grey circles). In SSL (right), pseudo-labels are generated from the data itself (striped circles) and used to train models to discriminate between samples. This illustration is inspired by a similar figure presented in [7]

was motivated by the realization that labelling them as purely unsupervised was somewhat misleading [5]. In contrast to traditional unsupervised learning, which typically aims to uncover latent structures in data without any explicit form of supervision, SSL relies on predictive tasks that generate an internal supervisory signal [6]. Although this signal is not derived from human-provided labels, it nonetheless guides learning by exploiting inherent relationships in the data. Hence, SSL has emerged as a distinct representation learning paradigm, distinct from both supervised and unsupervised learning. Figure 5.1 provides a conceptual illustration that highlights the differences between SSL and other representation learning paradigms.

Contrastive Learning for Time Series

Contrastive learning is a widely adopted SSL strategy that aims to learn meaningful representations by contrasting positive sample pairs against negative ones. The core idea is to construct a representation space in which positive pairs are pulled to be close together, while negative pairs are pushed farther apart [8, 9]. Following the recent success of contrastive learning in computer vision [10–13] and natural language processing [14, 15], increasing attention has been directed toward adapting these ideas to time series data. However, time series pose distinct challenges, including temporal dependencies, irregular sampling, and domain-specific semantics, which complicate the direct transfer of methods developed for other data modalities (e.g., images or text). Therefore, several methods have been proposed as baselines for contrastive learning in time series analysis.

Contrastive Predictive Coding (CPC) is designed to learn temporal representations by predicting future latent representations using an autoregressive model [16]. By maximizing mutual information between representations extracted from the same time series, CPC preserves temporal dependencies in the learned embeddings. Next, SimCLR generates positive pairs by applying data augmentations to the same input and applies a contrastive loss to maximize similarity between augmented views [17]. The key idea is to learn

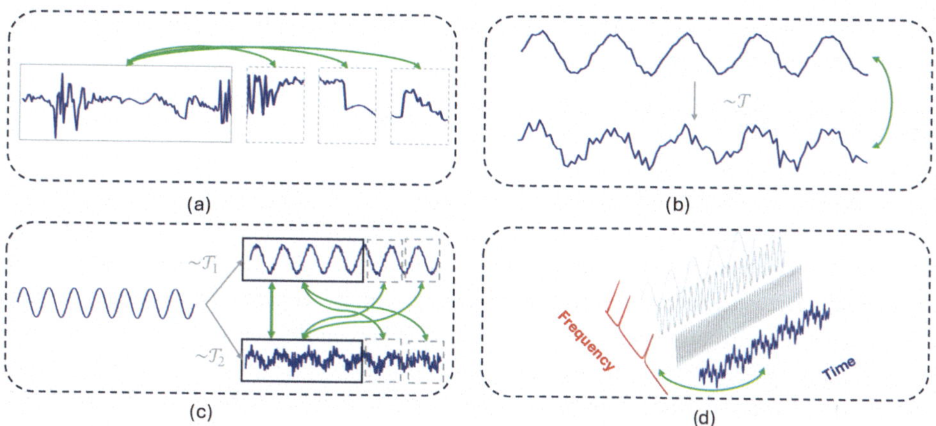

Fig. 5.2 Overview of positive pair selection strategies employed by representative contrastive learning methods for time series. (**a**) CPC: positive pairs are formed by contrasting past and future segments from the same signal; (**b**) SimCLR: positive pairs consist of the original signal and its augmented view; (**c**) TCC: positive pairs are constructed by contrasting contextual representations across augmented views and future segments; and (**d**) T-FC: positive pairs are formed between time-domain signals and their corresponding frequency-domain representations

augmentation-invariant representations, as differently augmented views of the same input have the same underlying semantic meaning. Although originally developed for image data, SimCLR has been adapted for time series, such as EEG signal analysis, by designing time series-specific augmentations [18]. Temporal Contextual Contrasting (TCC) combines a cross-view prediction task with a contextual contrastive objective to jointly learn robust temporal and discriminative features [19]. Time-Frequency Consistency (T-FC) further extends contrastive learning by leveraging spectral information, encouraging time-domain and frequency-domain representations of the same sample to be closer in a shared latent space than representations from different samples [20]. Despite their differences, these methods are primarily distinguished by their strategies for constructing contrastive pairs, with each approach employing a distinct sampling policy. Figure 5.2 summarizes the positive pair selection mechanisms used by these representative methods.

Although contrastive learning has shown promise for time series representation learning, several limitations remain. First, many existing approaches draw inspiration heavily from computer vision and natural language processing domains, where strong inductive biases such as invariance to transformations or cropping are commonly assumed [21]. These assumptions are not always valid for time series data; for example, cropping a time series can significantly alter its semantic meaning and statistical properties. Second, most methods focus on learning instance-level representations that summarize entire time series [21], which may be insufficient for tasks that require fine-grained or localized information, such as fault detection. Finally, recent contrastive learning approaches for time series often sample contrastive pairs exclusively along the temporal axis, neglecting the spectral dimension [22]. As a result, they fail to fully exploit the intrinsic time-frequency relationships present in many real-world time series.

Rationale for CDPCC

Predictive tasks that span multiple domains, such as the time and frequency domains, force models to learn representations that are informative and consistent across complementary views of the same data. For example, a time-domain signal and its frequency-domain representation describe the same underlying sample from different perspectives: the time domain emphasizes temporal ordering and local variations, whereas the frequency domain highlights periodic structures and spectral content. When model is learned to predict across these domains, we encourage the model to encode features that allow one representation to be inferred from the other. This cross-domain prediction mechanism can be viewed as a form of multi-view self-supervision, analogous to learning from multiple modalities of the same data. Consequently, the learned representations become more robust to domain-specific noise and capture variations that are shared across both domains.

Most existing time series contrastive learning methods rely primarily on intra-domain (same domain) comparisons, such as contrasting different augmented versions of a signal within the time domain only. This can facilitate learning of temporal features, but it may also lead to representations that overfit to time-specific patterns. Cross-domain prediction between time and frequency representations imposes a stronger constraint, as the model must learn features that map how patterns appear in both domains. This added difficulty can reveal latent structures that may remain hidden when analyzing a single domain in isolation. For instance, a transient spike in the time domain corresponds to a broad frequency spectrum.

SSL Evaluation Protocols

Consider a labeled time series dataset $\mathcal{D} = \{(X_1, Y_1), (X_2, Y_2), \ldots, (X_N, Y_N)\}$ consisting of N samples, where each input $X_i \in \mathbb{R}^{L\times p}$ comprises p variables over L time steps, and each label $Y_i \in \{0, \ldots, C-1\}$ corresponds to one of C possible classes. The objective is to learn a non-linear encoder g_E that maps each input X_i to a compact representation h_i that captures the underlying temporal and structural characteristics of the data.

During the self-supervised pre-training stage, the encoder g_E is optimized using a contrastive loss that exploits the intrinsic structure of the unlabeled inputs X, without access to class labels Y. Once pre-training is complete, the learned encoder parameters Θ are transferred to downstream tasks. Although the framework is applicable to general time series problems, this chapter focuses on fault detection scenarios, where X represents time-varying process measurements and Y denotes the corresponding operating condition (e.g., normal or faulty).

The quality and generalization capability of the learned representations h_i are assessed using three commonly adopted evaluation protocols, as illustrated in Fig. 5.3. In the linear evaluation protocol, the encoder parameters Θ are frozen, and a linear classifier is trained on top of g_E using labeled data from $\mathcal{D}$. In fine-tuning, both the encoder and the classifier

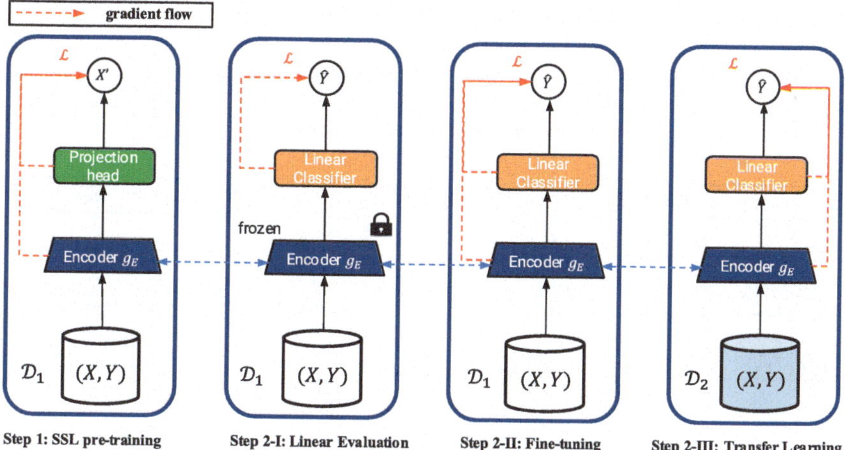

Fig. 5.3 Conceptual overview of SSL training and evaluation protocols. Step 1: Self-supervised pre-training, where the encoder is trained using a contrastive loss that exploits the inherent structure of X without relying on labels Y. A projection head is used only during pre-training and is discarded during evaluation. Step 2: Downstream evaluation using labeled data $\mathcal{D}$, following three common protocols: (I) Linear evaluation, where the encoder is frozen and only the linear classifier is trained; (II) Fine-tuning, where both the encoder and classifier are updated; and (III) Transfer learning, where the pre-trained model is adapted to a dataset different from that used during pre-training

are updated jointly using labeled data. In transfer learning (Step 2-III in Fig. 5.3), the pre-trained encoder and classifier are fine-tuned to a dataset that differs from the one used during self-supervised pre-training.

In this chapter, the quality of the learned representations is evaluated using the linear evaluation protocol (Step 2-I in Fig. 5.3).

5.2 CDPCC Framework

This section presents a description of the proposed CDPCC framework, which was originally introduced in [23]. An overview of the overall architecture is shown in Fig. 5.4. First, we slice the input time series into non-overlapping frames of equal length. Each frame is then transformed into the frequency domain using the fast Fourier transform (FFT) to obtain its spectral representation. Next, we introduce a cross-domain predictive contrasting module to model interactions between time- and frequency-domain dynamics using an autoregressive model. In this module, future embeddings in one domain are predicted using contextual information derived from the other domain. Moreover, a cross-domain contextual contrasting module is introduced to maximize agreement between context representations of the same sample across domains, thereby making the learned representations more discriminative. The source code for the proposed framework is

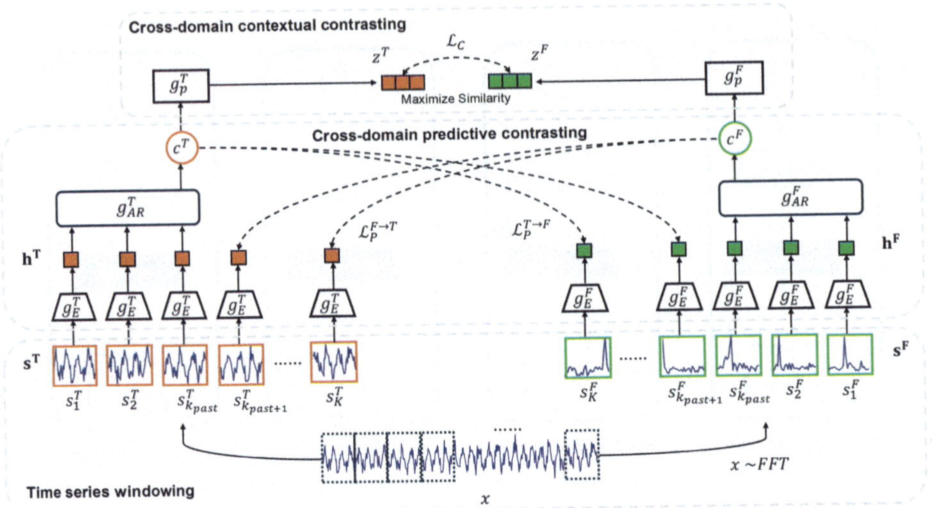

Fig. 5.4 Overall architecture of the proposed CDPCC framework. The model comprises six main components: time-domain and frequency-domain encoders (g_E^T and g_E^F), autoregressive models (g_{AR}^T and g_{AR}^F), and non-linear projection heads (g_P^T and g_P^F). The input time series X is sliced into K non-overlapping frames $\mathbf{s}^T$, each of which is transformed into its spectral counterpart $\mathbf{s}^F$ via FFT. Time-domain and frequency-domain representations are obtained as $\mathbf{h}^T = g_E^T(\mathbf{s}^T)$ and $\mathbf{h}^F = g_E^F(\mathbf{s}^F)$. Autoregressive models summarize the dynamics of the first k_{past} frames (with $k_{\text{past}} = 3$ in this illustration) to generate context vectors (c^T and c^F), which are then used to predict future embeddings in the opposite domain. A cross-domain contextual contrasting module further aligns these context vectors to learn discriminative feature representations

publicly available at https://github.com/iy641/CDPCC.git. The individual modules of the framework are described in detail in the following subsections.

5.2.1 Time Series Windowing

The CDPCC framework is designed to learn segment-level representations, where individual frames of a time series are encoded independently. Instance-level representations summarize an entire time series with a single embedding, whereas segment-level representations capture localized temporal patterns that may otherwise be overlooked in an instance-level approach. This design choice is particularly important for applications such as fault detection, where abnormal behaviour may occur over short time periods or exhibit small magnitudes. To facilitate segment-level learning, the input time series is split using a windowing strategy. Specifically, a time series X_i is divided into K non-overlapping frames of length d, yielding the set $\mathbf{s}_i^T = \{s_{i,1}^T, s_{i,2}^T, \ldots, s_{i,K}^T\}$, defined as

$$s_{i,k}^T = X_i[(k-1) \cdot d : k \cdot d], \quad 1 \leq k \leq K. \tag{5.1}$$

Contrastive learning aims to maximize agreement between different views of the same sample while minimizing similarity with representations of other samples. Traditional approaches typically generate multiple views through data augmentation. Selecting appropriate augmentation strategies that preserve semantic content is critical but often challenging, as such choices are highly task- and domain-dependent. Prior studies have shown that combining multiple augmentation techniques can improve downstream performance compared to relying on a single augmentation [17, 24]. However, time series data present additional challenges due to their intrinsic temporal structure. For instance, augmentations such as cropping or temporal shuffling may disrupt meaningful temporal dependencies.

In the CDPCC framework, rather than relying on handcrafted augmentations, it exploits the intrinsic relationship between the time and frequency domains as two complementary views of the same signal. The transformation between these domains is grounded in signal processing theory and provides a consistent form of invariance across different time series datasets. For each time-domain frame $s^T_{i,k}$, a Hamming window is first applied to prevent spectral leakage, followed by the computation of its FFT to obtain the corresponding frequency-domain representation $s^F_{i,k}$:

$$s^F_{i,k} = \mathrm{FFT}(s^T_{i,k} \odot w_h), \tag{5.2}$$

where w_h denotes the Hamming window function, defined as

$$w_h[n] = 0.54 - 0.46 \cos\left(\frac{2\pi n}{l-1}\right), \quad 0 \le n < l. \tag{5.3}$$

Since FFT produces complex-valued outputs, only the magnitude spectrum of the first $l/2$ frequency components is retained. This preserves the most informative spectral content while exploiting the symmetry property of the FFT output. This procedure produces the spectral representation $\mathbf{s}^F_i = \{s^F_{i,1}, s^F_{i,2}, \ldots, s^F_{i,K}\}$, which serves as a natural complementary view to the time-domain representation $\mathbf{s}^T_i$. Throughout this chapter, the superscripts ^{T} and ^{F} are used to denote time-domain and frequency-domain, respectively.

5.2.2 Cross-Domain Predictive Contrasting

The cross-domain predictive contrasting module is designed to capture dynamic relationships between time-domain and frequency-domain representations using a contrastive loss. After constructing paired views $\mathbf{s}^T_i$ and $\mathbf{s}^F_i$, each is processed by a domain-specific encoder to extract temporal and spectral features. The time-domain encoder g^T_E maps each frame $s^T_{i,k}$ to a latent embedding $h^T_{i,k} = g^T_E(s^T_{i,k})$, while the frequency-domain encoder g^F_E maps each spectral frame $s^F_{i,k}$ to its corresponding embedding $h^F_{i,k} = g^F_E(s^F_{i,k})$.

The resulting sequences of embeddings are denoted as $\mathbf{h}_i^T = \{h_{i,1}^T, h_{i,2}^T, \ldots, h_{i,K}^T\}$ and $\mathbf{h}_i^F = \{h_{i,1}^F, h_{i,2}^F, \ldots, h_{i,K}^F\}$, where each embedding lies in $\mathbb{R}^d$ $(d > p)$.

An autoregressive model g_{AR} is then used to summarize the dynamics of the first k_{past} frames in each domain. Given embeddings up to the $k_{\text{past}}^{\text{th}}$ frame, g_{AR} produces a context vector $c_i = g_{AR}(h_{i,k \le k_{\text{past}}})$, with $c_i \in \mathbb{R}^m$. The time-domain context vector c_i^T is used to predict future frequency-domain embeddings, while the frequency-domain context vector c_i^F is used to predict future time-domain embeddings.

Future embedding prediction is performed using a log-bilinear scoring function, following the formulation proposed in [16]. This score function measures how well a context vector from one domain predicts embeddings in the opposite domain. Separate score functions are defined for each prediction direction (time $\rightarrow$ frequency and frequency $\rightarrow$ time):

$$f_{i,k}^{T \rightarrow F}(c_i^T, h_{i,k}^F) = \exp\left((h_{i,k}^F)^\top \mathbf{W}_k^{T \rightarrow F} c_i^T\right), \quad k_{\text{past}} < k \le K, \tag{5.4}$$

$$f_{i,k}^{F \rightarrow T}(c_i^F, h_{i,k}^T) = \exp\left((h_{i,k}^T)^\top \mathbf{W}_k^{F \rightarrow T} c_i^F\right), \quad k_{\text{past}} < k \le K, \tag{5.5}$$

where $\mathbf{W}_k^{T \rightarrow F}$ and $\mathbf{W}_k^{F \rightarrow T}$ are learnable transformation matrices in $\mathbb{R}^{d \times m}$, defined separately for each prediction step k.

The objective of the cross-domain prediction task is to maximize agreement between predicted and true future embeddings of the same sample while minimizing similarity with embeddings from other samples in the batch $\mathcal{D}_B$. Accordingly, the predictive contrasting losses $\mathcal{L}_P^{T \rightarrow F}$ and $\mathcal{L}_P^{F \rightarrow T}$ are defined as

$$\mathcal{L}_P^{T \rightarrow F} = -\frac{1}{N_B(K - k_{\text{past}})} \sum_{i=1}^{N_B} \sum_{k=k_{\text{past}}+1}^{K} \log\left(\frac{\exp\left((h_{i,k}^F)^\top \mathbf{W}_k^{T \rightarrow F} c_i^T\right)}{\sum_{n \in \mathcal{D}_B} \exp\left((h_n^F)^\top \mathbf{W}_k^{T \rightarrow F} c_i^T\right)}\right), \tag{5.6}$$

$$\mathcal{L}_P^{F \rightarrow T} = -\frac{1}{N_B(K - k_{\text{past}})} \sum_{i=1}^{N_B} \sum_{k=k_{\text{past}}+1}^{K} \log\left(\frac{\exp\left((h_{i,k}^T)^\top \mathbf{W}_k^{F \rightarrow T} c_i^F\right)}{\sum_{n \in \mathcal{D}_B} \exp\left((h_n^T)^\top \mathbf{W}_k^{F \rightarrow T} c_i^F\right)}\right), \tag{5.7}$$

where N_B denotes the batch size.

5.2.3 Cross-Domain Contextual Contrasting

The cross-domain contextual contrasting module aims to learn discriminative representations by aligning context vectors from the time and frequency domains for the same sample in a shared (time-frequency) latent space. This module employ nonlinear projection heads, denoted g_P^T and g_P^F, which map the time-domain and frequency-domain context vectors c_i^T and c_i^F onto lower-dimensional representations z_i^T and z_i^F, respectively. Prior work has shown that including projection heads in contrastive learning frameworks can reduce computational overhead and improve the generalization of learned representations [17,25].

Given a batch containing N_B samples, two context representations are obtained per sample, yielding a total of $2N_B$ context vectors. The pair (z_i^T, z_i^F) corresponding to the same sample forms a positive pair, while the remaining $2N_B - 2$ context vectors from other samples constitute negative pairs. The objective is to maximize similarity between positive pairs and minimize similarity between negative pairs. The cross-domain contextual contrastive loss $\mathcal{L}_C$ is defined as

$$\mathcal{L}_C = -\frac{1}{N_B}\sum_{i=1}^{N_B} \log\left(\frac{\exp\left(\mathrm{sim}(z_i^T, z_i^F)/\tau\right)}{\sum_{j=1}^{2N_B} \mathbb{1}_{[i\neq j]} \exp\left(\mathrm{sim}(z_i^T, z_j^F)/\tau\right)}\right), \tag{5.8}$$

where $\mathrm{sim}(a, b) = \frac{a^\top b}{\|a\|\|b\|}$ denotes cosine similarity between normalized vectors, $\mathbb{1}_{[i\neq j]}$ is an indicator function equal to one when $i \neq j$, and τ is a temperature hyperparameter.

The overall CDPCC objective function combines the two predictive contrasting losses with the contextual contrasting loss:

$$\mathcal{L}_T = \lambda_1\left(\mathcal{L}_P^{T\to F} + \mathcal{L}_P^{F\to T}\right) + \lambda_2\mathcal{L}_C, \tag{5.9}$$

where λ_1 and λ_2 are scalar hyperparameters that control the relative contributions of the predictive and contextual loss terms.

5.3 Performance Evaluation on Benchmark Datasets

The proposed CDPCC framework is evaluated on two benchmark datasets, CSTH and Arc Loss, and compared against seven baseline approaches. These baselines include four state-of-the-art contrastive learning methods for time series: **SimCLR**, **CPC**, **TCC**, and **T-FC**. We also include **DTW+1NN**, a non-deep learning method that serves as a reference performance benchmark based on time series similarity. To examine the impact of self-supervised pre-training, two additional models are considered: (1) **Random Initialization (RI)**, where a linear classifier is trained on top of randomly initialized and frozen encoders g_E^T and g_E^F; and (2) **Supervised**, where the CDPCC architecture is trained end-to-end using labels. Both settings employ the same network architecture used in the contrastive CDPCC framework. This chapter focuses on time series contrastive learning methods rather than traditional fault-detection approaches, as their training objectives and evaluation protocols differ fundamentally.

We employ a hold-out strategy for model evaluation. A random search is conducted over a predefined hyperparameter space to identify well-performing configurations. Models that perform best on the validation set are then evaluated on the held-out test set, and results are reported on this final set. For both g_E^T and g_E^F, we use a 3-block 1D convolutional encoder followed by a non-linear projection layer. For the autoregressive models g_{AR}^T and g_{AR}^F, we use a standard two-layer LSTM architecture for simplicity. The projection heads g_P^T and g_P^F are implemented as two-layer fully connected networks without parameter sharing. We set $\tau = 0.2$, $\lambda_1 = 0.7$, and $\lambda_2 = 0.3$ in the loss function. All models are trained for 100 epochs with early stopping based on validation performance. Reported results correspond to the mean and standard deviation over five independent runs using the same data split. For DTW+1NN, the reported standard deviation is zero because the method is deterministic. The complete hyperparameter configurations for each dataset are provided in the corresponding configuration files (`/config_files/*`) within the GitHub repository.

5.3.1 Quantitative Evaluation: Model Prediction Performance

This subsection reports downstream fault detection results on CSTH and Arc Loss for the proposed CDPCC framework and all baseline approaches. A linear evaluation protocol is used to assess the quality of the representations learned during contrastive pre-training. In this setting, all parameters of the pre-trained encoder are frozen, and a new linear classifier is trained on top of the learned representations. The classifier is trained using the labelled training set and evaluated using five metrics. The results are summarized in Table 5.1.

The results in Table 5.1 show that CDPCC achieves strong performance relative to the baselines on both CSTH and Arc Loss. CDPCC ranks first on the Arc Loss benchmark and achieves near-top performance on CSTH, with results comparable to those of a fully supervised setting. Furthermore, CDPCC consistently outperforms DTW+1NN and RI baselines by a clear margin, which suggests that contrastive pre-training yields feature representations that transfer effectively to downstream fault detection.

Next, CDPCC also outperforms instance-level contrastive learning methods such as SimCLR and T-FC. Unlike methods that encode an entire time series into a single embedding, CDPCC learns segment-level representations that retain localized temporal variations, which are often important for fault detection. The overall performance gap between segment-level methods (e.g., CDPCC, CPC, and TCC) and instance-level methods (e.g., SimCLR and T-FC) suggests that local temporal structure provides more valuable information than purely global features for fault detection tasks.

Even though CPC, TCC, and CDPCC share a predictive contrastive component, in the sense that future embeddings are predicted from past information, CDPCC achieves better performance under the same evaluation protocol. Moreover, CPC and TCC perform same-domain (time-domain) predictions, while CDPCC performs cross-domain prediction between time- and frequency-domain embeddings. This cross-domain objective constrains

Table 5.1 Performance Comparison of CDPCC and baseline approaches on CSTH and Arc Loss datasets. Bold values indicate the best performance in each metric

Model	ACC (%)	PPV (%)	TPR (%)	1-FPR (%)	F_1 (%)
CSTH					
DTW+1NN	81.17 ± 0.00	96.38 ± 0.00	64.89 ± 0.00	97.55 ± 0.00	77.56 ± 0.00
RI	87.72 ± 0.74	90.71 ± 1.61	82.93 ± 1.72	92.13 ± 1.59	86.62 ± 0.83
Supervised	**99.55 ± 0.18**	**99.71 ± 0.25**	**99.40 ± 0.23**	**99.71 ± 0.25**	**99.55 ± 0.18**
SimCLR	82.42 ± 2.40	86.74 ± 2.72	76.37 ± 4.11	88.18 ± 2.67	81.16 ± 2.76
CPC	97.32 ± 0.60	98.69 ± 1.10	95.88 ± 0.84	98.71 ± 1.10	97.26 ± 0.60
TCC	97.25 ± 0.43	98.04 ± 0.74	96.08 ± 0.78	98.06 ± 0.74	97.05 ± 0.46
T-FC	95.59 ± 0.65	98.79 ± 0.22	92.07 ± 1.20	98.86 ± 0.19	95.31 ± 0.73
CDPCC	99.29 ± 0.18	99.53 ± 0.23	99.04 ± 0.30	99.53 ± 0.23	99.29 ± 0.18
Arc Loss					
DTW+1NN	64.96 ± 0.00	65.52 ± 0.00	64.31 ± 0.00	65.63 ± 0.00	64.91 ± 0.00
RI	68.13 ± 1.50	65.47 ± 1.40	78.57 ± 3.97	57.41 ± 3.88	71.37 ± 1.69
Supervised	75. 62 ± 0.60	71.86 ± 1.06	85.14 ± 3.06	65.93 ± 2.81	77.89 ± 0.88
SimCLR	66.28 ± 1.51	62.17 ± 0.85	83.93 ± 3.27	47.84 ± 1.42	71.41 ± 1.65
CPC	69.82 ± 1.12	69.13 ± 1.54	76.56 ± 7.45	64.76 ± 6.08	72.41 ± 2.45
TCC	74.88 ± 0.47	71.19 ± 2.32	**85.15 ± 5.82**	64.39 ± 6.21	77.35 ± 1.18
T-FC	69.29 ± 0.90	67.13 ± 0.55	77.79 ± 1.24	61.07 ± 0.80	72.06 ± 0.76
CDPCC	**75.88 ± 0.67**	**72.95 ± 1.64**	83.22 ± 3.68	**68.39 ± 3.92**	**77.96 ± 0.94**

the encoder to learn representations that remain informative across both domains. Furthermore, CDPCC surpasses T-FC, which aligns time and frequency representations but does not explicitly include a predictive objective. Incorporating prediction into the contrastive formulation appears to produce more robust representations, hence improving downstream fault detection performance.

Finally, SimCLR relies on manually selected augmentations to generate contrastive pairs. In time series, such transformations may not consistently preserve semantic meaning, which can adversely impact representation quality. On the other hand, CDPCC constructs contrastive pairs by leveraging the inherent relationship between time- and frequency-domain views. This strategy avoids the need for augmentation tuning and provides a more stable mechanism for generating positive pairs across different time series datasets.

5.3.2 Qualitative Evaluation: Comparative Analysis of Visual Representations

In this subsection, we analyze the encoded frame representations ($\mathbf{h}^T$ and $\mathbf{h}^F$) for the CSTH and Arc Loss datasets. Specifically, we explore the visual patterns contained within these representations to show how the time-domain and frequency-domain encoders, g_E^T and g_E^F, capture distinct yet complementary information.

The time-domain embeddings, $\mathbf{h}^T$, capture local temporal information that highlights changes in process dynamics over time. On the other hand, the frequency-domain embeddings, $\mathbf{h}^F$, extract spectral features that encode periodic components and frequency-based variations that may not be apparent in the time domain. We visualize the Hadamard product ($\mathbf{h}^T \circ \mathbf{h}^F$), the element-wise product of $\mathbf{h}^T$ and $\mathbf{h}^F$.

We adopt a qualitative approach to examine and interpret the visual representations across different process conditions in the benchmark datasets. The visualizations provide an intuitive understanding of how CDPCC learns meaningful representations and integrates information from both domains.

5.3.2.1 CSTH Benchmark

Figure 5.5 shows the visual representation of two CSTH samples recorded under normal operating conditions. The time series plots show smooth variations without abrupt changes or irregular fluctuations. The $\mathbf{h}^T \circ \mathbf{h}^F$ heatmaps exhibit uniform activation patterns across the sequential frames. This indicates that both encoders capture steady-state process characteristics.

Next, Fig. 5.6 presents two faulty CSTH samples with their corresponding visual representations. The left-hand side sample corresponds to random fluctuations in CW flow, while the right-hand side sample represents an abrupt pulse change introduced into the level transmitter. In the first sample, the visual representation shows a sharp increase in activation intensity in later segments, corresponding to the onset of oscillations. In contrast, the visual representation for the second sample exhibits distinct activation patterns.

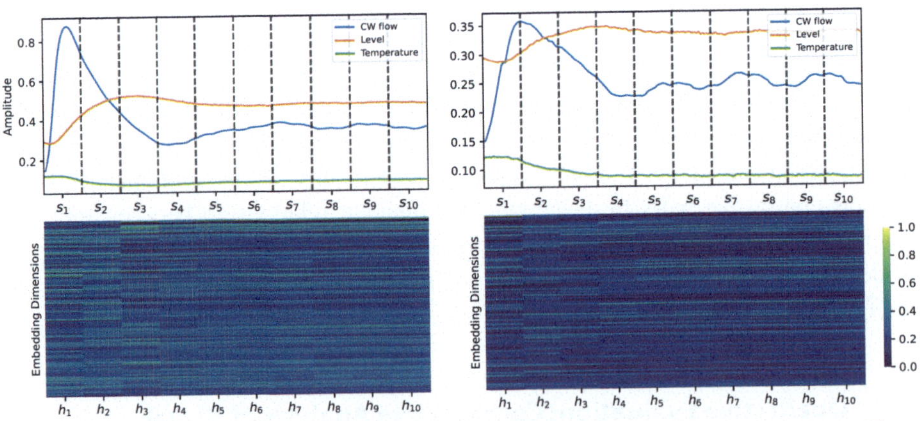

Fig. 5.5 Visual representations of two CSTH samples under normal operating conditions. The top row shows the time series plots. The process variables exhibit smooth variations without abrupt changes. The bottom row presents the corresponding Hadamard product heatmaps ($\mathbf{h}^T \circ \mathbf{h}^F$), where activation patterns remain uniform across embedding dimensions, which indicate stable and steady-state process characteristics

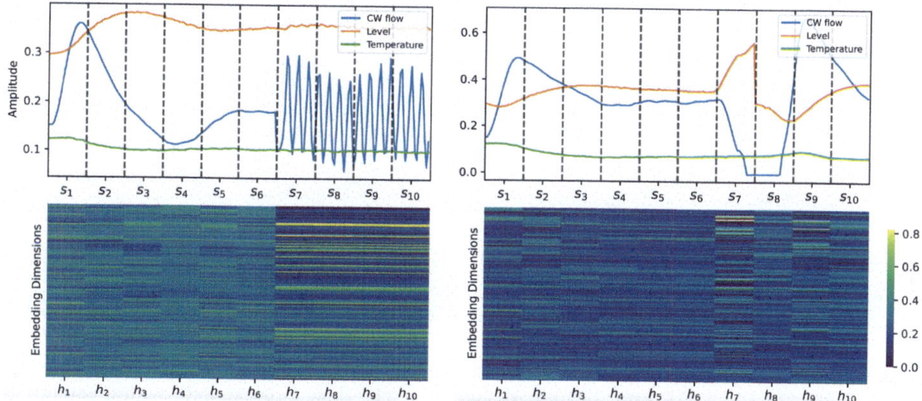

Fig. 5.6 Visual representations of two CSTH samples under faulty operating conditions. The top row shows the time series plots. The left-hand sample exhibits random oscillations in CW flow starting around s_7, while the right-hand sample shows an abrupt pulse change in the level transmitter around s_8. The bottom row presents the corresponding Hadamard product heatmaps ($\mathbf{h}^T \circ \mathbf{h}^F$), which highlight distinct activation patterns that align with the observed faults, particularly in the later segments for the left-hand sample and around s_8 for the right-hand sample

Specifically, the visual representation shows a high activation around s_8, corresponding to the moment or period of the fault (i.e., sudden change in the level transmitter signal).

5.3.2.2 Arc Loss Benchmark

We analyze the visual representations for the Arc Loss benchmark under normal and faulty operating conditions. Similar to the CSTH benchmark, we aim to interpret how the time-domain and frequency-domain encoders capture process information and highlight key differences between normal and faulty samples. Figure 5.7 displays two representative samples from the Arc Loss dataset under normal operating conditions. The top row shows the 3D time series plots of process variables, where the signals maintain relatively stable and consistent trends over time, with no significant deviations. The corresponding Hadamard product heatmaps ($\mathbf{h}^T \circ \mathbf{h}^F$) in the bottom row exhibit uniform activation patterns across all sequential frames. This consistency reflects the steady-state nature of the process during normal operation.

On the other hand, Fig. 5.8 illustrates the visual representations of two faulty samples from the Arc Loss benchmark. The left-hand sample initially exhibits stable behaviour up until s_7. Around this period, significant deviations in variables measurements are observed, which indicate the onset of an arc loss fault. However, the process is brought back to its normal operating conditions after the fault. This transition is reflected in the visual representation, where the activation patterns remain consistent before and after h_8. This highlights the temporary nature of the fault. Furthermore, the right-hand sample also begins under normal operating conditions but undergoes an abrupt change in process variables measurements due to a potential arc loss fault. Unlike the first sample, the

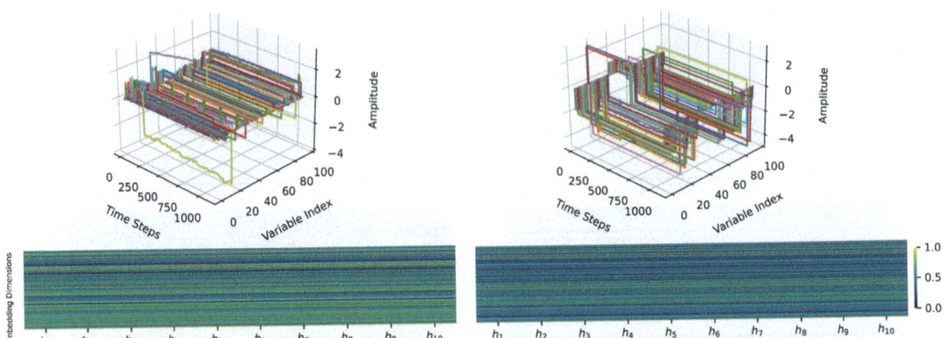

Fig. 5.7 Visual representations of two Arc Loss samples under normal operating conditions. The top row shows 3D time series plots of process variables. The time series signals exhibit smooth and stable behaviour. The bottom row presents the corresponding Hadamard product heatmaps ($\mathbf{h}^T \circ \mathbf{h}^F$), with uniform activation patterns indicating steady-state process conditions

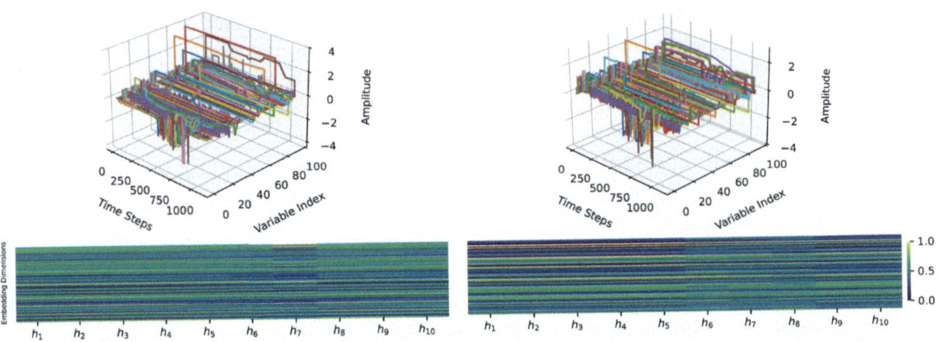

Fig. 5.8 Visual representations of two Arc Loss samples under faulty operating conditions. The top row shows 3D time series plots. In the left-hand sample, the process remains stable until around s_7, where significant deviations indicate an arc loss fault. The process is later stabilized; thus, similar activation patterns are observed before and after the fault. In the right-hand sample, an abrupt change occurs, marking an arc loss event. The process then stabilizes at a new steady state, which is reflected in the visual representation as a shift in activation patterns before and after the fault is observed

process does not return to its original steady state but instead stabilizes at a new operating condition. This shift is captured in the visual representation, where the activation patterns differ before and after the fault event.

5.3.3 Empirical Analysis

In this subsection, we examine the empirical performance of the CDPCC model under several practical settings. We first study its performance when labelled data are scarce, then investigate sensitivity to key hyperparameters, and finally assess the contribution of its individual components through an ablation study.

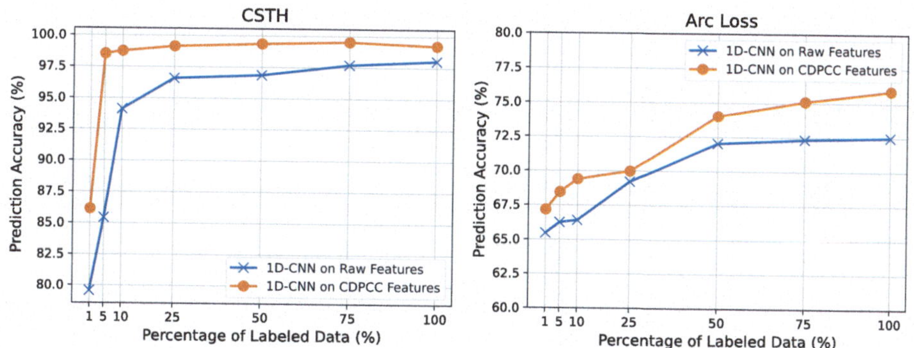

Fig. 5.9 Prediction accuracy comparison between 1D-CNN classifiers trained on raw inputs and on CDPCC-extracted features for the CSTH and Arc Loss datasets

5.3.3.1 Few-Labelled Data Scenarios

To evaluate CDPCC in low-label regimes, we consider two classifiers: one trained directly on raw input features and another trained on features extracted using the proposed CDPCC encoder. Both classifiers are trained using different fractions of the labelled training set, namely 1%, 5%, 10%, 25%, 50%, 75%, and 100%. In all cases, we employ a simple 3-block 1D-CNN as the downstream classifier.

We begin by reporting the performance of the supervised 1D-CNN trained on raw signals for CSTH and Arc Loss. On the CSTH benchmark, the classifier achieves 79.6% accuracy when trained with only 1% of the labelled data, and improves to 98.1% when trained using the full labelled set. On the Arc Loss benchmark, performance increases from 65.5% accuracy at 1% labels to 72.5% with 100% labels. These results correspond to the blue curve in Fig. 5.9.

Next, we assess whether CDPCC improves label efficiency by training the same 1D-CNN classifier on features produced by the CDPCC model, rather than on raw inputs. In this setting, the CDPCC feature extractor remains fixed, and only the downstream 1D-CNN classifier is trained until convergence. The resulting performance is shown by the orange curve in Fig. 5.9.

We observe that using CDPCC-extracted features yields consistent accuracy gains across label fractions. For example, with 1% of labelled data, the CDPCC-based classifier achieves 86.2% accuracy on CSTH and 67.2% on Arc Loss, corresponding to improvements of 6.6% and 1.7% over the supervised baseline, respectively. CDPCC also outperforms when trained with the full labelled set. The results indicate improved data efficiency: with 50% labelled data, the CDPCC-based classifier surpasses the supervised baseline trained on 100% of the labels for both datasets. Similarly, under the 1% label setting, the CDPCC-based classifier outperforms the baseline trained with 5% labels, corresponding to approximately a $5\times$ improvement in label efficiency.

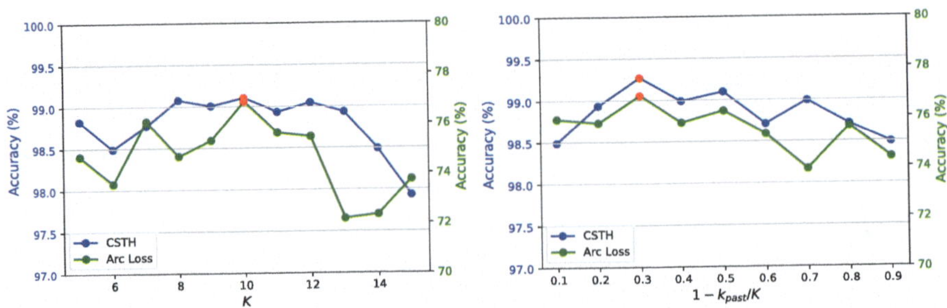

Fig. 5.10 Sensitivity analysis results on CSTH (blue) and Arc Loss (green). Left: effect of the total number of frames K on prediction accuracy. Right: effect of the predicted-future fraction $(1 - k_{past}/K)$ on prediction accuracy. Red markers indicate the highest accuracy achieved for each dataset

5.3.3.2 Sensitivity Analysis

Here, we conduct sensitivity experiments to analyze the influence of two key hyperparameters: the total number of frames K and the number of future frames predicted, $(K - k_{past})$. Here, K is determined by the time series windowing module, while k_{past} controls the amount of past context provided to the autoregressive model in the cross-domain predictive contrasting module.

First, we evaluate prediction performance on CSTH and Arc Loss as K varies from 5 to 15. Smaller values of K correspond to longer frames (i.e., more time steps per frame). The left panel of Fig. 5.10 summarizes the effect of K on accuracy. The results suggest that increasing K can reduce performance, plausibly because each frame contains less information and the frequency representation becomes more susceptible to spectral leakage. In our experiments, $K = 10$ provides the strongest overall performance and is therefore used in subsequent evaluations.

Second, we analyze the impact of k_{past} by varying the proportion of future frames relative to the total frames, expressed as $(1-k_{past}/K)$. The right panel of Fig. 5.10 reports the corresponding accuracy trends. Generally, predicting a larger portion of future frames improves performance up to a point. However, predicting too many future frames can degrade performance because it leaves insufficient past context to train the autoregressive model. The best results are obtained when predicting approximately 30% of the frames; accordingly, we set $(1 - k_{past}/K)$ to 30% in our experiments (i.e., $k_{past} = 0.7K$).

5.3.3.3 Ablation Study

We conduct an ablation study on the CSTH dataset to quantify the contribution of individual CDPCC components. The full CDPCC model is compared to three variants: (1) *w/o predictive contrasting*, which removes the cross-domain predictive losses ($\mathcal{L}_P^{T\rightarrow F}$ and $\mathcal{L}_P^{F\rightarrow T}$); (2) *w/o contextual contrasting*, which removes the cross-domain contextual loss ($\mathcal{L}_C$); and (3) *w/o projection layers*, which removes the projection heads (g_P^T and g_P^F) and computes the contrastive loss directly on the context vectors (c^T and c^F) instead

Table 5.2 Ablation results on the CSTH dataset

	Avg. accuracy
CDPCC	**99.29%**
w/o Predictive contrasting ($\lambda_1 = 0$)	97.70% (−1.59%)
w/o Contextual contrasting ($\lambda_2 = 0$)	98.45% (−0.84%)
w/o Projection layers (g_P^T and g_P^F)	98.39% (−0.90%)
Autoregressive model g_{AR} architectures	
LSTM	**99.29%**
→ GRU	98.49% (−0.80%)
→ Transformer	86.95% (−12.34%)

of their projected versions (z^T and z^F). The results are reported in Table 5.2. Across all cases, removing any component leads to a clear reduction in accuracy, indicating that each module contributes meaningfully to the overall performance of CDPCC.

We also examine the choice of autoregressive architecture by replacing the LSTM with a GRU and a Transformer model of comparable parameter size. Both alternatives lead to a noticeable drop in accuracy, suggesting that the LSTM provides a more suitable autoregressive backbone for the proposed framework in this setting.

References

1. Balestriero R, Ibrahim M, Sobal V, Morcos A, Shekhar S, Goldstein T, Bordes F, Bardes A, Mialon G, Tian Y, Schwarzschild A, Wilson AG, Geiping J, Garrido Q, Fernandez P, Bar A, Pirsiavash H, LeCun Y, Goldblum M. A cookbook of self-supervised learning. *arXiv preprint arXiv:2304.12210*. 2023. https://arxiv.org/abs/2304.12210
2. Caron M, Bojanowski P, Joulin A, Douze M. Deep clustering for unsupervised learning of visual features. *arXiv preprint arXiv:1807.05520*. 2019.
3. Wu Z, Xiong Y, Yu S, Lin D. Unsupervised feature learning via non-parametric instance-level discrimination. *arXiv preprint arXiv:1805.01978*. 2018. https://arxiv.org/abs/1805.01978
4. Doersch C, Gupta A, Efros AA. Unsupervised visual representation learning by context prediction. In: *Proceedings of the IEEE International Conference on Computer Vision (ICCV)*. 2015. p. 1422–1430. https://doi.org/10.1109/ICCV.2015.167
5. LeCun Y, Misra I. Self-supervised learning: The dark matter of intelligence. Facebook AI Blog. 2020. https://ai.facebook.com/blog/self-supervised-learning-the-dark-matter-of-intelligence/
6. Deldari S, Xue H, Saeed A, He J, Smith DV, Salim FD. Beyond just vision: A review on self-supervised representation learning on multimodal and temporal data. *arXiv preprint arXiv:2206.02353*. 2022.
7. Schmarje L, Santarossa M, Schröder S-M, Koch R. A survey on semi-, self- and unsupervised learning for image classification. *IEEE Access*. 2021;9:82146–82168.
8. Le-Khac PH, Healy G, Smeaton AF. Contrastive representation learning: A framework and review. *IEEE Access*. 2020;8:193907–193934. https://doi.org/10.1109/ACCESS.2020.3031549
9. Kumar P, Rawat P, Chauhan S. Contrastive self-supervised learning: Review, progress, challenges and future research directions. *International Journal of Multimedia Information Retrieval*. 2022;11(4):461–488. https://doi.org/10.1007/s13735-022-00245-6

10. He K, Fan H, Wu Y, Xie S, Girshick R. Momentum contrast for unsupervised visual representation learning. *arXiv preprint arXiv:1911.05722*. 2020.
11. Chen X, Fan H, Girshick R, He K. Improved baselines with momentum contrastive learning. *arXiv preprint arXiv:2003.04297*. 2020.
12. Grill J-B, Strub F, Altché F, Tallec C, Richemond PH, Buchatskaya E, Doersch C, Avila Pires B, Guo ZD, Azar MG, Piot B, Kavukcuoglu K, Munos R, Valko M. Bootstrap your own latent: A new approach to self-supervised learning. *arXiv preprint arXiv:2006.07733*. 2020.
13. Caron M, Misra I, Mairal J, Goyal P, Bojanowski P, Joulin A. Unsupervised learning of visual features by contrasting cluster assignments. *arXiv preprint arXiv:2006.09882*. 2021.
14. Mikolov T, Chen K, Corrado G, Dean J. Efficient estimation of word representations in vector space. *arXiv preprint arXiv:1301.3781*. 2013. https://arxiv.org/abs/1301.3781
15. Reimers N, Gurevych I. Sentence-BERT: Sentence embeddings using Siamese BERT-networks. *arXiv preprint arXiv:1908.10084*. 2019. https://arxiv.org/abs/1908.10084
16. van den Oord A, Li Y, Vinyals O. Representation learning with contrastive predictive coding. *arXiv preprint arXiv:1807.03748*. 2018.
17. Chen T, Kornblith S, Norouzi M, Hinton G. A simple framework for contrastive learning of visual representations. *arXiv preprint arXiv:2002.05709*. 2020. https://arxiv.org/abs/2002.05709
18. Falck F, Sarkar SK, Roy S, Hyland SL. Contrastive representation learning for electroencephalogram classification. In: *ML4H@NeurIPS*. 2020.
19. Eldele E, Ragab M, Chen Z, Wu M, Kwoh CK, Li X, Guan C. Time-series representation learning via temporal and contextual contrasting. *arXiv preprint arXiv:2106.14112*. 2021. https://arxiv.org/abs/2106.14112
20. Zhang X, Zhao Z, Tsiligkaridis T, Zitnik M. Self-supervised contrastive pre-training for time series via time-frequency consistency. *arXiv preprint arXiv:2206.08496*. 2022. https://arxiv.org/abs/2206.08496
21. Yue Z, Wang Y, Duan J, Yang T, Huang C, Tong Y, Xu B. TS2Vec: Towards universal representation of time series. *Proceedings of the AAAI Conference on Artificial Intelligence*. 2021.
22. Yang L, Qiao L. Unsupervised time-series representation learning with iterative bilinear temporal-spectral fusion. *Proceedings of the International Conference on Machine Learning*. 2022.
23. Yousef I, Shah SL, Gopaluni RB. Time series representation learning via cross-domain predictive and contextual contrasting: Application to fault detection. *Engineering Applications of Artificial Intelligence*. 2025;154:110914. https://doi.org/10.1016/j.engappai.2025.110914
24. Um TT, Pfister FMJ, Pichler D, Endo S, Lang M, Hirche S, Fietzek U, Kulić D. Data augmentation of wearable sensor data for Parkinson's disease monitoring using convolutional neural networks. In: *Proceedings of the 19th ACM International Conference on Multimodal Interaction (ICMI '17)*. 2017. p. 216–220. https://doi.org/10.1145/3136755.3136817
25. Gupta K, Ajanthan T, van den Hengel A, Gould S. Understanding and improving the role of projection head in self-supervised learning. *arXiv preprint arXiv:2212.11491*. 2022. https://arxiv.org/abs/2212.11491

6 Conclusion

The increasing complexity of modern industrial systems presents both opportunities and challenges. For example, there are potential gains that can be leveraged for improved performance and robustness. However, the sources of faults and disturbances are numerous and often unforeseen, making it difficult to predict or diagnose them. Furthermore, the growing demands for operational efficiency, environmental compliance, and just-in-time operations amplify the consequences of process faults, as any deviation can lead to significant economic, environmental, and societal repercussions. Therefore, ensuring the smooth and reliable operation of such systems has become a critical objective for both engineers and researchers.

To address the aforementioned challenges, advanced process data analytics and machine learning have emerged as powerful tools. These technologies offer the means to analyze large-scale industrial data, facilitating the early detection of faults, accurate fault diagnosis, and actionable insights for operational improvements. By learning complex patterns within the data, machine learning models can meet the dynamic and evolving needs of modern industrial systems. However, successfully applying machine learning to such problems often requires more than just powerful algorithms. It necessitates a deep understanding of the data itself and innovative approaches to represent it in a way that optimizes interpretability and downstream performance.

In industrial settings, process data are predominantly in the form of time series, which rank among the most challenging data types to analyze. Time series data often involve multiple variables recorded over time, which makes them inherently high-dimensional and difficult to visualize. Moreover, the temporal dependencies within the data must be preserved, as these relationships often hold the key to understanding system behaviour. These challenges make time series data particularly difficult to model using traditional machine learning techniques. To bridge this gap, this monograph proposes a novel approach: visual analytics, which transforms time series data into visual representations

I. Yousef et al., *Visual Analytics for Process Monitoring*, Synthesis Lectures on Emerging Engineering Technologies, https://doi.org/10.1007/978-3-032-22126-1_6

to uncover patterns that are not readily apparent in the raw time domain while preserving the temporal dependencies within the data.

The transformation of time series data into visual representations serves multiple purposes. First, it facilitates the identification of latent patterns and relationships by exploiting spatial and visual characteristics that are more interpretable to both humans and algorithms. Second, this approach preserves the temporal dependencies inherent in the data. This ensures that system dynamics are not lost in the transformation process. Third, visual representations reduce the complexity of high-dimensional data, which makes it more accessible for both automated analysis and manual inspection. Fourth, this transformation allows us to leverage the maximum capabilities of advanced computer vision models (e.g., CNN), which have demonstrated exceptional performance in extracting complex features and achieving state-of-the-art results in a variety of domains. This monograph addresses a bottleneck in the application of machine learning to industrial systems by focusing on converting process data into interpretable visual forms.

In future work, a key direction is to extend the visual analytics framework from offline to online training for real-time industrial FDD. The current work relies on historical process data for model training, but practical deployments must continuously adapt to new data. This transition introduces challenges like data drift and concept shifts: as process operating conditions evolve, the statistical properties of data streams may change suddenly or incrementally. Future research should focus on online learning algorithms that update incrementally under computational constraints (limited retraining time and memory) while maintaining model stability and performance. This online adaptation (e.g. sliding windows or weighted updates) is critical for timely fault detection, which allows the visual analytic tools to promptly adjust to new operating regimes, ultimately improving the robustness and accuracy of process monitoring.

Another promising future research direction is to leverage the bijective mapping functions (time series $\leftrightarrow$ images) for data cleaning. We seek to answer the following research question: "Would data cleaning (e.g., denoising or imputing missing values) using the encoded images yield more stable and interpretable results than directly preprocessing raw time series?". Encoding time series data into two-dimensional representations (e.g., using GAF) allows the use of powerful image processing techniques on temporal data. In the image domain, noise can be reduced, and missing data can be "inpainted" using well-established filters or neural inpainting methods. This might preserve complex temporal patterns that might be distorted by naive time-domain interpolation. Image-based preprocessing pipelines could provide structured and intuitive ways to clean time series data. For instance, outliers or anomalies might appear as distinct visual patterns that can be smoothed, and consistent trends or periodicities may be easier to extract in a 2D format. Reverting the preprocessed images back to time series form ensures that any data preprocessing is retained in the original time domain.

Finally, another future research direction includes examining whether encoding time series as images is truly the optimal representation for machine learning tasks. Encoding time series into images in this work offers several advantages: it enables the use of powerful computer vision models, and provides a data format that humans can intuitively interpret (e.g. seeing a fault pattern as a distinctive texture or shape). However, other representational paradigms are emerging. An intriguing possibility is to transform time series into a textual or symbolic sequence, thereby enabling the application of large language models (LLMs) for FDD analysis. Building on this idea, future models might encode temporal patterns into a linguistic form; for example, by quantizing signal behaviours into tokens or phrases, and feed these sequences into advanced LLMs. This could open a new pathway for interpretable analysis.

The Arc Loss Dataset: List of Variables

A

The list of variables for the Arc Loss dataset is shown in Table A1.

Table A1 List of Arc Loss process variables

Tag	Description	Range	Unit
Progress parameters			
t	Time [yyyy-mm-dd hh:mm:ss]		
Smelting parameters			
AP	Electrode A power	0–60	MW
BP	Electrode B power	0–60	MW
TP	Total power (AP+BP)	0–100	MW
APSP	Electrode A power set point	0–60	MW
BPSP	Electrode B power set point	0–60	MW
AC	Electrode A current	0–100	kA
BC	Electrode B current	0–100	kA
ACSP	Electrode A current set point	0–100	kA
BCSP	Electrode B current set point	0–100	kA
AV	Electrode A voltage	<2200	Volts
BV	Electrode B voltage	<2200	Volts
AVSP	Electrode A voltage set point	≤ 2200	Volts
BVSP	Electrode B voltage set point	≤ 2200	Volts
AR	Resistance around electrode A	≥ 0	mΩ
BR	Resistance around electrode B	≥ 0	mΩ
ARSP	Resistance around electrode A set point	≥ 0	mΩ
BRSP	Resistance around electrode B set point	≥ 0	mΩ

(continued)

I. Yousef et al., *Visual Analytics for Process Monitoring*, Synthesis Lectures on Emerging Engineering Technologies, https://doi.org/10.1007/978-3-032-22126-1

Table A1 (continued)

Tag	Description	Range	Unit
SER	Specific Energy Ratio	410–440	W/ton
AL	Arc A length		mm
BL	Arc B length		mm
CrucHL	Crucible (the wall) heat loss	0–5	MW
RoofHL	Roof heat loss	0–5	MW
PCHL	Plain cooler heat loss	≥0	MW
UWCHL	Upper chilled water heat loss	≥0	MW
LWCHL	Lower chilled water heat loss	≥0	MW
HFansHL	Hearth (Fans) heat loss	≥0	MW
HTCsHL	Hearth (Technological Control System) heat loss	≥0	MW
SL	Slag level		mm
ML	Metal level		mm
FOGET	Off-gas temperature	180–630	°C
TPA	Slag tap A valve opening	0–100	%
TPB	Slag tap B valve opening	0–100	%
CO2	CO_2 volume	0–25	%
ST	Slag temperature after being tapped		°C
Furnace feed parameters			
FF	Furnace feed rate	0–200	tons/hr
FFDiv5APos	Furnace Feed Inlet Diverter 5A Position 1	[0,1]	
FFDiv5BPos	Furnace Feed Inlet Diverter 5B Position 1	[0,1]	
FFDiv5CPos	Furnace Feed Inlet Diverter 5C Position 1	[0,1]	
FFDiv5DPos	Furnace Feed Inlet Diverter 5D Position 1	[0,1]	
FFDiv5EPos	Furnace Feed Inlet Diverter 5E Position 1	[0,1]	
FFDiv5FPos	Furnace Feed Inlet Diverter 5F Position 1	[0,1]	
W1OC	Weir 1 valve opening	[0,1]	
W2OC	Weir 2 valve opening	[0,1]	
W3OC	Weir 3 valve opening	[0,1]	
W4OC	Weir 4 valve opening	[0,1]	
W5OC	Weir 5 valve opening	[0,1]	
W6OC	Weir 6 valve opening	[0,1]	
W7OC	Weir 7 valve opening	[0,1]	
W8OC	Weir 8 valve opening	[0,1]	
W1F	Weir 1 flow rate	0–200	tons/hr
W2F	Weir 2 flow rate	0–200	tons/hr
W3F	Weir 3 flow rate	0–200	tons/hr
W6F	Weir 6 flow rate	0–200	tons/hr

(continued)

Table A1 (continued)

Tag	Description	Range	Unit
W1	Weir 1 flow rate	≥0	mA
W2	Weir 2 flow rate	≥0	mA
W3	Weir 3 flow rate	≥0	mA
W4	Weir 4 flow rate	≥0	mA
W5	Weir 5 flow rate	≥0	mA
W6	Weir 6 flow rate	≥0	mA
W7	Weir 7 flow rate	≥0	mA
W8	Weir 8 flow rate	≥0	mA
TMF	Total microwave flow rate	≥0	mA
W1T	Feed temperature after leaving weir 1		°C
W2T	Feed temperature after leaving weir 2		°C
W3T	Feed temperature after leaving weir 3		°C
W4T	Feed temperature after leaving weir 4		°C
W5T	Feed temperature after leaving weir 5		°C
W6T	Feed temperature after leaving weir 6		°C
W7T	Feed temperature after leaving weir 7		°C
W8T	Feed temperature after leaving weir 8		°C
IT1	Feed temperature in the distribution bin 1		°C
IT2	Feed temperature in the distribution bin 2		°C
IT3	Feed temperature in the distribution bin 3		°C
IT4	Feed temperature in the distribution bin 4		°C
PAT	Feed temperature entering port A		°C
PB1T	Feed temperature entering port B1		°C
PB2T	Feed temperature entering port B2		°C
PB3T	Feed temperature entering port B3		°C
PB4T	Feed temperature entering port B4		°C
PB5T	Feed temperature entering port B5		°C
PB6T	Feed temperature entering port B6		°C
PC1T	Feed temperature entering port C1		°C
PC2T	Feed temperature entering port C2		°C
PC3T	Feed temperature entering port C3		°C
PC4T	Feed temperature entering port C4		°C
PC5T	Feed temperature entering port C5		°C
PC6T	Feed temperature entering port C6		°C
PC7T	Feed temperature entering port C7		°C
PC8T	Feed temperature entering port C8		°C

(continued)

Table A1 (continued)

Tag	Description	Range	Unit
PC9T	Feed temperature entering port C9		°C
PC10T	Feed temperature entering port C10		°C
WBCVPF	Weigh bin cone valve position feedback	≥ 0	mA
WBCVOMV	Weigh Bin Cone valve opening (Measured Value)	0–100	%
WBCVPCO	Weigh Bin cone valve position controller output	≥ 0	mA
CVPCSP	Cone valve position controller set point	0–100	%
FFPAPAF	Furnace feed pipe A, A-port flow	≥ 0	tons/hr
Reduction parameters			
FBRLSP	Fluidized bed reducer level set point		m
FBRL	Fluidized bed reducer level		m
FBRCMV	Fluidized bed reducer level		mA
FBRLCO	Fluidized bed reducer level controller output	≥ 0	mA
FBRCVC	Fluidized bed reducer cone valve control	0–100	%
FBRBedT	Fluidized bed reducer temperature	800–1100	°C
Calcining parameters			
CalcFR	Calciner feed rate	≥ 0	tons/hr
CoalFeed	Coal feed rate	≥ 0	tons/hr
Laboratory parameters			
AL2O3	Al_2O_3 concentration in the slag		ppm
FeO	FeO concentration in the slag		ppm
MgO	MgO concentration in the slag		ppm
Ni	Ni concentration in the slag		ppm
SiO2	SiO_2 concentration in the slag		ppm

The manufacturer's authorised representative in the EU is Springer Nature Customer Service Centre GmbH, Europaplatz 3, 69115 Heidelberg, Germany. If you have any concerns regarding our products, please contact ProductSafety@springernature.com

Printed and bound by CPI Group (UK) Ltd, Croydon, CR0 4YY
07/07/2026
02160926-0005